12/78

Q
162
A8.14

OF MATTERS GREAT
AND SMALL

ESSAYS ON SCIENCE by Isaac Asimov

From *The Magazine of Fantasy and Science Fiction*

Fact and Fancy
View from a Height
Adding a Dimension
Of Time and Space and Other Things
From Earth to Heaven
Science, Numbers, and I
The Solar System and Back
The Stars in Their Courses
The Left Hand of the Electron
The Tragedy of the Moon
Asimov on Astronomy
Asimov on Chemistry
Of Matters Great and Small

From Other Sources

Only a Trillion
Is Anyone There?
Today and Tomorrow and . . .

ISAAC ASIMOV

OF MATTERS GREAT
AND SMALL

DOUBLEDAY & COMPANY, INC.

GARDEN CITY, NEW YORK

1975

The following essays in this volume are reprinted from *The Magazine of Fantasy and Science Fiction*, having appeared in the following issues:

Constant as the Northern Star	August 1973
Signs of the Times	September 1973
The Mispronounced Metal	October 1973
The Figure of the Fastest	November 1973
The Figure of the Farthest	December 1973
The Eclipse and I	January 1974
Dance of the Luminaries	February 1974
The Uneternal Atoms	March 1974
A Particular Matter	April 1974
At Closest Range	May 1974
The Double-Ended Candle	June 1974
As Easy As Two Plus Three	July 1974
Updating the Asteroids	August 1974
Look Long upon a Monkey	September 1974
O Keen-eyed Peerer into the Future	October 1974
Skewered!	November 1974

© 1973, 1974 by Mercury Press, Inc.

In addition, one essay included in this volume appeared in *Science Digest:*

The Inevitability of Life	June 1974

© 1974 by The Hearst Corp.

Library of Congress Cataloging in Publication Data

Asimov, Isaac, 1920–
    Of matters great and small.

    CONTENTS: Stars: The figure of the fastest. The figure of the farthest. Constant as the northern star. Signs of the times.—Solar system: The eclipse and I. Dance of the luminaries. Updating the asteroids.—Life: The inevitability of life. Look long upon a monkey. O keen-eyed peerer into the future. [etc.]
    1. Science   2. Astronomy.   I. Title.
Q162.A814   508′.1
ISBN 0-385-02225-5
Library of Congress Catalog Card Number 74–12673

Dedicated to:
  Ed and Audrey Ferman,
    for co-operation above and beyond

Contents

Introduction

I am a creature of habit. I suppose that any prolific writer *must* be or he cannot be prolific. As soon as you fail to cultivate that automatic shuffle toward the typewriter immediately after breakfast and begin to say, "Oh, well, let's have a little variety. What *else* can I do today?" you're through. You may stay a writer in your own eyes, but you will become a mere dilettante.

Fortunately, this doesn't happen to me. If you were to take away my various typewriters during the night, I would sit down in my chair the next morning and twiddle my fingers in mid-air for five minutes before I realized that nothing was happening.

This business of "creature of habit" extends to the various *details* of the writing business, too, and nothing is so ingrained a part of my way of life as my monthly essay on science for *The Magazine of Fantasy and Science Fiction.*

It is a matter of gratification, I believe, to the Gentle Editor of that periodical (to whom this book is dedicated, by the way) that I have never even come close to missing a deadline for nearly two hundred consecutive issues.

I try to wear a virtuous expression when I meet him as a result, but actually, there's no virtue in it. Desperately I try not to get too far ahead of deadline. (The article I sent off today was mailed three months and four days ahead of deadline, as my resolution crumbled utterly, and Ed Ferman is going to have to call upon all the Gentleness in his soul to keep from telling me that there is such a thing as carrying punctuality too far. Of course, I did apologize humbly in the covering letter.)

In fact, every part of my science essays has grown stylized by habit; and any attempt to change that now would bring about a grinding of internal gears that might well damage my interior workings.

For instance, the articles are virtually never shorter than 3,900 words or longer than 4,300. I don't even have to pace myself or watch the page numbers. Something inside me winds up to the proper pitch of tension and then winds down as I type, and that's it.

Then, too, there is the matter of the personal anecdote that invariably starts the essay. When the series first began, in 1958, I never dreamed of doing such a thing. I simply started each essay with the subject at hand. Sometimes, of course, I felt that since the essays appeared in a science fiction magazine, I ought to justify them by tying them in, when I could, with science fiction. This meant, almost inevitably, that I would have to talk about myself.

By the time 1960 had come and I had written a dozen or so of the essays, I was talking about myself in the first few paragraphs every time; and this has continued ever since.

In fact, this evil practice broadened out and infected other parts of my writing. When I prepared my first anthology, *The Hugo Winners* (Doubleday, 1962), I found I had to do the introduction my own way. Uncertainly, I brought that introduction to my then-editor, Timothy Seldes, and said, "Listen, how does this sound?" I began reading and here's the way it started:

"Let me introduce this book my own way, please; by which I mean I will begin by introducing myself. I am Isaac Asimov and I am an old-timer. Not, you understand, that I am (ha, ha) really old. Quite the contrary. I am rather young actually, being only mumblety-mumble years old and looking even younger!"

By that time, Tim's assistant, the incredibly beautiful Wendy Weil, caught her breath and said, "You're making that up! You didn't really write it!"

But I had written it and what's more, after a bit of hesitation, Doubleday published it as written and confirmed me in my evil habits. My own story collections began to get autobiographical introductions, until, as in *The Early Asimov* (Doubleday, 1972) and *Before the Golden Age* (Doubleday, 1974), such collections became full-fledged autobiographies.

As you might expect, what started as a kind of personal touch, a kind of stylistic signature, has gotten out of control. It's not just that I've grown used to it; my *readers* have grown used to it. If I can judge from the letters that reach me, they are pleased with it; it removes some of the impersonality of the printed page and they feel my existence more clearly. It also means that they feel perfectly free to write to me and correct all my errors—but I value that, frankly. There's scarcely an article I've ever written that hasn't been corrected or amplified at one point or another by a genially smiling Friendly Reader, and you have no idea how grateful I am. It's something no amount of money can buy.

So when people ask me why I talk about myself so much in the course of my writing and hint that it is the result of a swinishly besotted love affair I have with myself, I always answer austerely that I do so because my readers insist on it.

Here's a case in point—

Mr. Richard Dempewolff, the jovial editor of *Science Digest* (for which I also write a monthly column, but one much shorter and quite different from my *F & SF* essays) asked me for a full-length article and I agreed.

But, having written it, I found, rather to my amazement, that I had hooked up my fingers to the wrong brain-outlet (or whatever it is I do when I sit down to write in a specific manner for a specific audience) and had turned out an *F & SF* essay. It was exactly the right length and exactly the right tone, it seemed to me.

Goodness, I thought (for in the privacy of my mind I use strong expletives), does that mean Dick won't take it?

I thought, Oh, well, if he doesn't like this, I'll write something else for him and send this one to Ed Ferman. (I am nothing if not philosophical about literary misadventures.)

However, as it turned out, Dick called me up and said he was taking the article. "Right down the center of the alley, Isaac," he said.

So I thought, Well, heck (what did I tell you about expletives?), why not use it anyway? I'll put it in my next collection.

And I have. It's Chapter 8, "The Inevitability of Life." The only thing is that, as written for *Science Digest*, it didn't have an introductory personal anecdote, and I didn't think my readers would sit still for its absence. So I wrote one especially for this collection.

You wouldn't have wanted me to omit it, would you?

No, I thought not.

ISAAC ASIMOV
New York City

OF MATTERS GREAT
AND SMALL

A—STARS

1. The Figure of the Fastest

As you can all imagine, I frequently receive outlines of odd theories invented by some of my readers. Most of them deal with vast concepts like the basic laws underlying all of space and time. Most of them are unreadable (or over my head, if you prefer.) Many of them are produced by earnest teen-agers, some by retired engineers. These theorists appear to think I possess some special ability to weigh deep and subtle concepts, combined with the imagination not to be deterred by the wildly creative.

It is all, of course, useless. I am no judge of great, new theories. All I can do is send back the material (which sometimes extends to many pages and forces me to incur substantial expense in postage) and try to explain, humbly, that I cannot help them.

Once in a while, though—once in all too long a while—I get a letter that I find amusing. One such came years ago. It was in fourteen vituperative, increasingly incoherent, pages of prose which boiled down to a diatribe against Albert Einstein, one that came under two headings:

1. Albert Einstein had gained world renown (my correspondent said) through the advancement of a great and subtle theory of relativity which he had stolen from some poor hard-working scientist. Einstein's victim thereupon died in obscurity and neglect without ever receiving the appreciation he deserved for this monumental discovery.

2. Albert Einstein had gained world renown (my correspondent also said) by inventing a completely false and ridiculous theory of relativity which had been foisted on the world by a conspiracy of physicists.

My correspondent argued *both* theses alternately with equal vehemence and clearly never saw that they were incompatible. Naturally, I didn't answer.

But what is there that causes some people to react so violently against the theory of relativity? Most of the people who object (usually much more rationally than my unfortunate correspondent, of course) know very little about the theory. About the only thing they know (and all that almost any non-physicist knows) is that according to the theory, nothing can go faster than light, and that offends them.

I won't go into the question of why scientists believed that nothing possessing mass can go faster than light, having handled that in several articles in the past.

I would, however, like to talk about the actual speed limit, the speed of light, what it actually is and how that was determined.

Olaus Roemer, the Danish astronomer, was the first to advance a reasonable figure for the speed of light through a study of the eclipses of Jupiter's satellites by Jupiter.

In 1676 he estimated that it took light 22 minutes to cross the extreme width of Earth's orbit about the Sun. At that time the total width of Earth's orbit was thought to be in the neighborhood of 174,-000,000 miles, so Roemer's results implied a speed of light of 132,-000 miles per second.

That is not bad. The figure is roughly 30 per cent low, but it is in the right ball park and for a first effort it is quite respectable. Roemer at least determined, correctly, the first figure of the value. The speed of light is indeed between 100,000 and 200,000 miles per second.

The next measurement of the speed of light came about, quite accidentally, a half century later.

The English astronomer James Bradley was trying to detect the parallax (that is, tiny shifts in position) of the nearer stars relative to the farther ones. This shift would result from the change in the position of the Earth as it moved around the Sun.

Ideally, every star in the heaven should move in an ellipse in the course of one year, the size and shape of that ellipse depending on the distance of the star from the Sun and its position with respect to the plane of Earth's orbit.

The farther the star, the smaller the ellipse and for all but the nearest stars the ellipse would be too small to measure. Those farther stars could therefore be considered motionless, and the displacement of the nearer stars relative to them would be the parallax Bradley was looking for.

Bradley *did* detect displacements of stars, but they were not what would be expected if Earth's motion around the Sun were responsible. The displacements could not be caused by parallax but had to be caused by something else. In 1728 he was on a pleasure sail on the Thames River and noted that the pennant on top of the mast changed direction according to the relative motion of ship and wind and *not* according to the direction of the wind alone.

That set him to thinking— Suppose you are standing still in a rainstorm with all the raindrops falling vertically downward because there is no wind. If you have an umbrella, you hold it directly over your head and remain dry. If you are walking, however, you will walk into some raindrops that have just cleared the umbrella if you continue to hold the umbrella directly over your head. You must angle the umbrella a little in the direction you are walking if you want to remain dry.

The faster you walk or the slower the raindrops fall, the farther you must tilt your umbrella to avoid walking into the raindrops. The exact angle through which you must tilt your umbrella depends on the ratio of the two velocities, that of the raindrops and that of yourself.

The situation is similar in astronomy. Light is falling on the Earth from some star in some direction and at some velocity. Meanwhile the Earth is moving around the Sun at another velocity. The telescope, like the umbrella, cannot be aimed directly at the star to gather the light but must be tilted a little in the direction the Earth is moving. (This is called "the aberration of light.") Because light is traveling very much faster than the Earth is moving in its orbit, the velocity ratio is high and the telescope must be tilted only very slightly indeed.

The tilt can be measured and, from that, the ratio of the speed of light to the speed of Earth in its orbit can be calculated. Since the Earth's orbital speed was known with fair accuracy, the speed of light could be calculated. Bradley calculated that that speed was such that light would cross the full width of Earth's orbit in 16 minutes 26 seconds.

If the width of Earth's orbit were 174,000,000 miles, this meant that light must travel at a rate of about 176,000 miles per second. This second try at the determination of the speed was considerably higher than Roemer's and considerably closer to the figure we now accept. It was still nearly 5 per cent low, however.

The methods of Roemer and Bradley both involved astronomical observations and had the disadvantage of depending for its accuracy on knowledge concerning the distance of the Earth from the Sun. This knowledge was still not very precise even through the nineteenth century. (If the width of the orbit had been known as accurately in Bradley's time as it is now, his figure for the speed of light would have been within 1.6 per cent of what we now consider it to be.)

Was it possible, then, to devise some method for measuring the speed of light directly by Earthbound experiments? In that case, the shakiness of astronomical statistics would be irrelevant. But how? Measuring a velocity that seems to be not too far below 200,000 miles per second presents a delicate problem.

In 1849 a French physicist, Armand Hippolyte Louis Fizeau, devised a way to turn the trick. He placed a light source on a hilltop and a mirror on another hilltop 5 miles away. Light flashed from the source to the mirror and back, a total distance of 10 miles, and it was Fizeau's intention to measure the time lapse. Since that time lapse was sure to be less than 1/10,000 of a second, Fizeau couldn't very well use a wrist watch, and he didn't.

What he did do was to place a toothed disc in front of the light source. If he held the disc motionless, the light would shoot out between two adjacent teeth, reach the mirror, and be reflected back between the teeth.

Suppose the disc were set to rotating. Light would travel so quickly that it would be at the mirror and back before the space between the teeth would have a chance to move out of the way. But now speed up the rate of rotation of the disc. At some speed the light ray would flash to the mirror and back only to find that the disc had turned sufficiently to move a tooth in the way. The reflected light ray could no longer be observed.

Make the disc move still more rapidly. The light ray would then flash outward between two teeth and be reflected back at a time when the tooth had moved past and the *next* gap was in the path of the light ray. You could see its reflection again.

If you knew how rapidly the disc rotated, you would know the fraction of a second it would take for a tooth to move in the way of the reflected ray and how long for that tooth to move out of the way of the reflected ray. You would then know how much time it took light to cover 10 miles and, therefore, how far it would go in a second.

The value Fizeau settled on turned out to be about 196,000 miles per second. This was no better than Bradley's value and was still 5 per cent off, but it was now too high rather than too low.

Helping Fizeau in his experiments was another French physicist, Jean Bernard Léon Foucault. Foucault eventually went on to attempt to measure the speed of light on his own, according to a slightly different type of experiment.

In Foucault's scheme, the light still flashed from a source to a mirror and then back. Foucault arranged it, however, so that on its return, the light ray fell on a second mirror, which reflected the ray onto a screen.

Suppose, now, you set the second mirror to revolving. When the light returns, it hits the second mirror after it has changed its angle just slightly and the light ray is then reflected on the screen in a slightly different place than it would if the second mirror had been motionless.

Foucault set up the experiment in such a way that he was able to measure this displacement of the light ray. From this displacement and knowing how fast the second mirror was revolving, Foucault could calculate the speed of light.

Foucault's best measurement, made in 1862, was about 185,000 miles per second. This was the most nearly accurate measurement yet made. It was only 0.7 per cent low and Foucault was the first to get the second figure correct. The speed of light was indeed somewhere between 180,000 and 190,000 miles per second.

Foucault's measurement was so delicate that he didn't even have to use particularly great distances. He didn't use adjacent hilltops but carried out the whole thing in a laboratory with a light ray that traveled a total distance of about 66 feet.

The use of such a short distance led to something else. If light is expected to travel 10 miles it is very difficult to have it travel through anything but air or some other gas. A liquid or solid may be transparent in short lengths but 10 miles of any liquid or solid is simply opaque. Over a distance of 66 feet, however, it is possible to make a beam of light shine through water or through any of a variety of other media.

Foucault passed light through water and found that by his method its velocity was considerably slower, only three fourths of its velocity in air. It turned out in fact that the speed of light depended on the index of refraction of the medium it traveled through. The higher the index of refraction, the lower the speed of light.

But air itself has an index of refraction, too, though a very small one. Therefore, the speed of light, as measured by Fizeau and Foucault, had to be a trifle too low no matter how perfect the measurement. In order to get the maximum speed of light one would have to measure it in a vacuum.

As it happens, the astronomical methods of Roemer and Bradley involved the passage of light through the vacuum of interplanetary and interstellar space. The light in each case also passed through the full height of the atmosphere but that length was insignificant compared to the millions of miles of vacuum the light had crossed. However, the astronomical methods of the eighteenth and nineteenth centuries had sources of error that utterly swamped the tiny advantage inherent in substituting vacuum for air.

The next important figure in the determination of the speed of light was the German-American physicist Albert Abraham Michelson. He began working on the problem in 1878 by using Foucault's scheme but improving the accuracy considerably. Whereas Foucault had to work with a displacement of the spot of light of only a little over 1/40 of an inch, Michelson managed to produce a displacement of some 5 inches.

In 1879 he reported the speed of light to be 186,355 miles per second. This value is only 0.04 per cent too high and was by far the most accurate yet obtained. Michelson was the first to get the third figure right, for the speed of light was indeed between 186,000 and 187,000 miles per second

Michelson kept working, using every possible way of increasing the precision of the measurement, especially since, by 1905, Einstein's theory of relativity made the speed of light seem a fundamental constant of the Universe.

In 1923 Michelson picked two mountaintops in California, two that were not 5 miles apart as Fizeau's had been, but 22 miles apart. He surveyed the distance between them till he had that down to the nearest inch! He used a special eight-sided revolving mirror and by 1927 announced that the speed of light was about 186,295 miles per second. This was only 0.007 per cent too high and now he had the first four figures correct. The speed of light was indeed between 186,200 and 186,300 miles per second.

Michelson still wasn't satisfied. He wanted the speed of light *in a vacuum*. It was that speed and nothing else that was a fundamental constant of the Universe.

Michelson therefore used a long tube of accurately known length and evacuated it. Within it, he set up a system that sent light back and forth in that tube till he had made it pass through 10 miles of vacuum. Over and over he made his measurements, and it wasn't till 1933 that the final figure was announced (two years after he had died.)

The final figure was 186,271 miles per second and that was a small further approach to the truth, for it was only 0.006 per cent too low.

In the four decades since Michelson's final determination, physicists have developed a variety of new techniques and instruments which might be applied to the determination of the speed of light.

For instance, it became possible to produce light of a single wavelength by means of a laser beam and to measure that wavelength to a high degree of precision. It was also possible to determine the frequency of the wavelength (the number of oscillations per second) with equally high precision.

If you multiply the length of one wavelength by the number of wavelengths per second, the product is the distance covered by light in one second—in other words, the speed of light.

This was done with greater and greater precision, and in October 1972 by far the most accurate measurement ever made was announced by a research team headed by Kenneth M. Evenson, working with a chain of laser beams at the National Bureau of Standards laboratories in Boulder, Colorado.

The speed they announced was 186,282.3959 miles per second.

The accuracy of the measurement is within a yard in either direction so, since there are 1,760 yards in a mile, we can say that the speed of light is somewhere between 327,857,015 and 327,857,017 yards per second.

Of course, I have been giving all the measurements in common units of miles, yards, and so on. Despite all my scientific training, I still can't visualize measurements in the metric system. It's the fault of the stupid education all American children get—but that's another story.

Still, if I don't think in the metric system instinctively, I can at least handle it mathematically and I intend to use it more and more in these essays. The proper way to give the speed of light is not in miles per second or in yards per second, but in kilometers per second and in meters per second. Using the proper language, the speed of light is now set at 299,792.4562 kilometers per second. If we multiply it by 1,000 (the beauty of the metric system is that so many multiplications and divisions are so simple), it is equal to 299,792,456.2 meters per second, give or take a meter.

There are few measurements we can make that are as accurate as the present value of the speed of light. One of them is the length of the year which is, in fact, known with even greater precision.

Since the number of seconds in a year is 31,556,925.9747, we can calculate the length of a light-year (the distance light will travel in one year) as 5,878,499,776,000 miles, or 9,460,563,614,000 kilometers. (There's no use trying to figure out that final 000. Even now the speed of light is not accurately enough known to give the light-year to closer than a thousand miles or so.)

All these figures are, of course, un-round and are troublesome to memorize exactly. This is too bad since the speed of light is so fundamental a quantity, but it is to be expected. The various units—miles, kilometers, and seconds—were all determined for reasons that had nothing to do with the speed of light and therefore it is in the highest degree unlikely that that speed would come out even. That we even come near a round figure is merely a highly fortunate coincidence.

In miles per second, the common value given for the speed of light in, let us say, a newspaper story, is 186,000 miles per second, which is only 0.15 per cent low. This is good enough, but there are three figures that must be memorized—186.

In kilometers per second we have a much better situation, since if we say the speed of light is 300,000 kilometers per second, we are only 0.07 per cent low. The approximation is twice as close as in the miles-per-second case, and only one figure need be remembered, the 3. (Of course, you must also remember the order of magnitude— that the speed is in the hundreds of thousands of kilometers per second and not in the tens of thousands or in the millions.)

The beauty of the metric system again displays itself. The fact that the speed of light is about 300,000 kilometers per second means that it is about 300,000,000 meters per second and about 30,000,000,000 centimeters per second, all three figures being at the same approximation to the truth.

If we use exponential figures, we can say that the speed of light is 3×10^5 kilometers per second, or 3×10^8 meters per second, or 3×10^{10} centimeters per second. You need only memorize one of these since the others are easily calculated from the one, provided you understand the metric system. The exponential figure 10^{10} is particularly easy to remember, so if you associate that with "centimeters per second" and then don't forget to multiply it by 3, you've got it made.

The fact that the speed of light is so close to a pretty round number in the metric system is, of course, a coincidence. Let's locate that coincidence.

One of the most convenient measures of distance that people use is the distance from the nose to the tip of the fingers of an arm stretched horizontally away from the body. You can imagine someone selling a length of textile or rope or anything flexible by stretching out successive lengths in this manner. Consequently, almost every culture has some common unit of about this length. In the Anglo-American culture it is the "yard."

When the French Revolutionary committee were preparing a new system of measurements in the 1790s, they needed a fundamental unit of length to begin with and it was natural to choose one that would approximate the good old nose-to-fingertip length. To make it non-anthropocentric, however, they wanted to tie it to some natural measurement.

In the previous decades, as it happened, Frenchmen had taken the lead in two expeditions designed to make exact measurements of the curvature of the Earth in order to see if it were flattened at the poles, as Isaac Newton had predicted. That placed the exact size and shape of the Earth very much in the consciousness of French intellectuals.

The Earth proved to be slightly flattened, so the circumference of the Earth passing through both poles was somewhat less than the circumference around the Equator. It seemed very up to date to recognize this by tying the fundamental unit of length to one of these particularly. The polar circumference was chosen because one of these could be made to go through Paris, whereas the equatorial circumference (the one and only) certainly did not go through that city of light.

By the measurements of the time, the polar circumference was roughly equal to 44,000,000 yards, and that quadrant of the circumference from the Equator to the North Pole, passing through Paris, was about 11,000,000 yards long. It was decided to make the length of the quadrant just 10,000,000 times the fundamental unit and to define the new unit as 1/10,000,000 of that quadrant and give it the name of "meter."

This definition of the meter was romantic but foolish, for it implied that the polar circumference was known with great precision which, of course, it was not. As better measurements of the Earth's vital statistics were made, it turned out that the quadrant was very slightly longer than had been thought. The length of the meter could not be adjusted to suit—too many measurements had already been made with it; and the quadrant is now known to be *not* 10,000,000 meters long as it ought to be by French logic, but 10,002,288.3 meters long.

Of course, the meter is no longer tied to the Earth. It was eventually defined as the distance between two marks on a platinum-iridium rod kept with great care in a vault at constant temperature and, finally, as so many wavelengths of a particular ray of light (the orange-red light emitted by the noble gas isotope krypton-86, to be exact).

Now for the coincidences:

1. It so happens that the speed of light is very close to 648,000 times as great as the speed at which the Earth's surface at the

Equator moves as our planet rotates on its axis. This is just a coincidence, for the Earth could be rotating at any velocity and was in the past rotating considerably faster and will in the future be rotating considerably slower.

2. A single rotation of the Earth is defined as a day and our short units of time are based on exact divisions of the day. Thanks to the Babylonians and their predecessors, we use the factors 24 and 60 in dividing the day into smaller units and by coincidence 24 and 60 are also factors of 648,000. As a result of coincidences 1 and 2, anything moving at the speed of light will make a complete circle at Earth's Equator almost exactly 450 times per minute, or almost exactly 7.5 times per second—which are simple numbers.

3. Since, by a third coincidence, the French commissioners decided to tie the meter to the circumference of the Earth and make it an even fraction of that circumference, the result is an inevitable near-round number for the speed of light in the metric system. There are 40,000,000 meters (roughly) to Earth's circumference and if you multiply this by 7.5 you come out with 300,000,000 meters per second.

Can we do better? Can we just have an exponential figure without having to multiply it? Can we express the speed as a certain number of units of length per unit of time with a number that consists of a 1 followed by a number of zeros and come fairly close to the truth?

If we multiply 3 by 36 we come out with a product of 108. If we remember that there are 3,600 seconds in the hour, it follows that the speed of light is 1,079,252,842 kilometers per hour. This is just about 8 per cent over the figure of 1,000,000,000 kilometers per hour. If we were to say that the speed of light is 10^9 kilometers per hour, we'd be only 8 per cent low of the facts and that's not too bad, I suppose.

As for the light-year, we can say it is 6,000,000,000,000 (six trillion) miles and be only 2 per cent high. To express that exponentially, however, we must say 6×10^{12} miles and that multiplication by 6 is a nuisance. In the metric system we can say that a light-year is ten trillion kilometers, or 10^{13} kilometers, and be only 6 per cent high. The lesser accuracy might be more than counterbalanced by the elegance of the simple figure 10^{13}.

However, honesty compels me to say that the despised common measurements happen to offer a closer way of approaching the light-year in a purely exponential way. If we say the light-year is equal to 10^{16} yards, we are only 3.5 per cent high.

2. The Figure of the Farthest

I sometimes despair of people ever getting anything right. From personal experience I have grown doubtful about trusting even the best histories and biographies. They may be right in the grand sweep, but it doesn't seem possible to get the little details as they really were.

For instance, I do nothing but talk about myself in almost everything I write, so you would think there would be some details about my personal life that would be well-known to anyone who is interested in me and in my writing. Well, not so!

I once received a copy of the April 29, 1973, issue of *Silhouette Magazine*, published by the Colorado Springs *Sun*. In it, there is an article on science fiction that includes a (telephone) interview with me. Aside from a few typographical errors, it is a very nice article and I am very pleased with it—except for one sentence.

The article quotes a Mr. Clayton Balch who, it says, teaches two science fiction courses at El Paso Community College. Mr. Balch talks, in part, about the drug culture and its influence on science fiction. Apparently he thinks that writers need some sort of artificial stimulation and simply adopt whatever variety is handy in their time. The article quotes Mr. Balch as saying about the drug culture, "A lot of the younger writers grew up with it, and in the same way Asimov is drinking scotch, younger writers are using drugs."

Well, darn it, Asimov does not use drugs and is NOT drinking scotch either, and never did. Asimov is a teetotaler and has said so in print at least fifty times and has demonstrated it in public at least a million times. And yet, in the future (if there is one), biographers, combing every last bit of mention about me, will come across this item and solemnly record that scotch was my favorite drink. (Actually, I do like a little sip of a sweet wine like Manischewitz

Concord grape, or Cherry Heering, or even Bristol Cream sherry— but even a little sip gets me high, so it's not really a good idea to try.)

If a simple little thing like my drinking habits can't be straightened out, it's no wonder that more subtle difficulties offer a great deal of trouble. For instance, although the situation has been explained in a million astronomy books, and in several of my own articles, I am continually bombarded with letters from people who are idignant at the fact that galaxies are receding from us at a rate proportional to their distance from us. "What is so special about *us?*" they insistently ask.

In the past I have explained that this recession in proportion to distance (Hubble's Law) can be accounted for by the expansion of the Universe, but I have never really explained in detail. Now I will, because I've thought of a way of doing it that I've never seen anyone else try.

—But I won't get to that right away. I will sneak up on it in my usual oblique fashion by making the article deal, first, with the successive enlargements of man's picture of the Universe.

To begin with, men only knew the size of that portion of the Universe with which they made direct contact and this, generally, wasn't much. Traders and generals, however, were bound to travel great distances as the ancient empires grew in size.

In 500 B.C., when the Persian Empire stretched from India to Egypt over an extreme width of 3,000 miles, Hecataeus of Miletus, the first scientific geographer among the Greeks, estimated the land surface of the Earth (which he considered to be flat) to be a circular slab about 5,000 miles in diameter. This, then, is our first figure for the longest straight line that was more or less accurately known:

1. 500 B.C.—5,000 miles

By 350 B.C. the Greek philosophers were quite certain that the Earth was a sphere, and about 225 B.C., Eratosthenes of Cyrene, noting that sunlight hit different portions of the Earth's surface at different angles at the same time, used the fact to calculate the size of that sphere. He worked it out correctly, making the Earth's diameter 8,000 miles, and this became the longest known straight line:

2. 225 B.C.—8,000 miles

But the Earth's diameter could be no final maximum, since beyond the Earth lay the heavenly bodies. About 150 B.C. Hipparchus of Nicaea, the greatest of all Greek astronomers, calculated the distance to the Moon by valid trigonometric methods and announced that distance to be equal to thirty times the diameter of the Earth. Accepting Eratosthenes' figure for that diameter, we get the distance to the Moon to be about 240,000 miles, which is correct. If we imagine a sphere centered on the Earth and large enough to contain the Moon's orbit, its diameter is 480,000 miles, and that becomes the maximum straight line accurately measured:

3. 150 B.C.—480,000 miles

And the other heavenly bodies? Between Hecataeus and Hipparchus, the known size of the Universe had increased ninety-six-fold. It had doubled in measured size every fifty years, on the average. Could this not have continued? At that rate, the distance to the Sun could have been determined about A.D. 250.

Not so, alas. After Hipparchus there came an eighteen-century dead halt. To use trigonometric methods for determining the distance of objects farther from the Earth than the Moon required a telescope, and that was not invented until 1608.

In 1609 the German astronomer Johannes Kepler first worked out the model of the solar system, but it was not until 1671 that the first reasonably accurate parallax of a planet (Mars) was made, telescopically, by the Italian-French astronomer Giovanni Domenico Cassini.

Using that parallax and Kepler's model, Cassini worked out the distances of the various bodies of the solar system from the Sun. His figures were about 6 per cent low by contemporary standards, but I'll ignore such first-time inaccuracies in measurements made by valid methods and use the correct figures. Thus, Saturn, which was the farthest planet known in Cassini's time, is 886,000,000 miles from the Sun. If we imagine a sphere centered on the Sun and large enough to include Saturn's orbit, its diameter would be the new longest accurately measured length:

4. 1671—1,800,000,000 miles

This was nearly four thousand times the length of greatest distance accurately known to the ancients and shows the power of the telescope.

It did not remain a record long, however. In 1704 the English astronomer Edmund Halley worked out the orbit of Halley's Comet, and it seemed to him that it receded to a distance of 3,000,000,000 miles from the Sun before returning. On the basis of his calculations he predicted the return of the comet, and its return in 1758 (the year he had predicted) proved him right. The diameter of a sphere centered on the Sun and including the orbit of Halley's Comet was the new record:

5. 1704—6,000,000,000 miles

However, all the astronomers working in the first two centuries of the telescopic era knew that measuring the distances within the solar system would in no way tell them the size of the Universe. Outside the solar system were the stars.

Astronomers worked hard attempting to determine the distances of the stars by measuring their extremely small parallaxes, and in the 1830s three astronomers succeeded, almost simultaneously.

The German astronomer Friedrich Wilhelm Bessel announced the distance of the star 61 Cygni in 1838. The Scottish astronomer Thomas Henderson announced the distance of Alpha Centauri in 1839, and the German-Russian astronomer Friedrich Wilhelm von Struve announced the distance of Vega in 1840.

Of these, Vega was the most distant, being about 160,000,000,000, 000 miles from here. These are too many zeros to handle conveniently. By the 1830s some fairly good estimates already existed for the speed of light, so that it was possible to use the "light-year" as a unit of distance; that is, the distance that light would travel in one year. This comes out to be about 5,880,000,000,000 (see Chapter 1) so that Vega is about 27 light-years distant. If we take a sphere, then, which is centered on the Sun and is large enough to contain Vega, its diameter would be the new record distance:

6. 1840—320,000,000,000,000 miles, or 54 light-years

This was an enormous fifty-thousand-fold increase over solar system distances, but it could be no record, for beyond Vega lay uncounted other and more distant stars. As early as 1784 the German-English astronomer William Herschel had counted the stars in different directions, to see if they extended outward symmetrically. They didn't, and Herschel was the first to suggest that the system of stars existed as a flattened lens-shaped object which we now call the "Galaxy."

Herschel tried to estimate the size of the Galaxy, but could produce only a very hazy guess. In 1906, however, a Dutch astronomer, Jacobus Cornelis Kapteyn, knowing the distance to the nearer stars and having at his disposal huge star maps and the new technique of photography, estimated that the long-diameter of the Galaxy was 55,000 light-years:

7. 1906—55,000 light-years

This represented a thousandfold increase over the period of the first discovery of stellar distances, but it was not yet enough. By 1920 the American astronomer Harlow Shapley, making use of the period of Cepheid variables as a new way of determining distances, showed that the Galaxy was much larger than Kapteyn had thought. (The figure, using Shapley's methods, is now thought to be 100,000 light-years.) In addition, Shapley could show that the Magellanic Clouds were systems of stars lying just outside the Milky Way and were up to 165,000 light-years from us. A sphere centered on the Sun, and large enough to include the Magellanic Clouds, would have a diameter that would set a new record of length:

8. 1920—330,000 light-years

This was a sixfold increase over Kapteyn's figure and did it represent, at last, the entire Universe? There were many astronomers, even as late as 1920, who suspected that the Galaxy and the Magellanic Clouds were all there was to the Universe and that beyond them lay nothing.

There was, however, considerable doubt about the Andromeda nebula, a cloudy patch of whiteness which some thought was a mere patch of luminous fog within the Galaxy and others thought lay far outside the Galaxy and, indeed, was another galaxy as large as our own. The matter was not finally settled until 1923, when the

American astronomer Edwin Powell Hubble made out individual stars in the outskirts of the nebula and was able to determine its distance. He showed that it was far outside the Galaxy and was certainly a galaxy in its own right. Twenty years later, the method he used was modified and the distance of the Andromeda galaxy turned out to be four times as far as Hubble had first thought.

If we imagine a sphere centered on the Sun and including the Andromeda galaxy (using the distance-figure of 2,700,000 light-years now accepted) we have the diameter of that sphere as the new record:

9. 1923—5,400,000 light-years

This sixteenfold increase over Shapley's figure, however, brought a new humility in its train, for once again it was clear that the new record wasn't much of a record. Once the Andromeda was recognized as a galaxy, it was at once realized that millions of other and dimmer patches of luminous fog must also be galaxies and that all of them were farther than the Andromeda galaxy was.

Through the 1920s and 1930s the distances of dimmer and dimmer galaxies were determined by studying the characteristics of their spectra. By 1940 men like the American astronomer Milton La Salle Humason had found galaxies that were as far distant as 200,000,000 light-years. A sphere centered on the Sun and enclosing them would supply a diameter for a new record:

10. 1940—400,000,000 light-years

This seventy-five-fold increase over the distance of the Andromeda galaxy did not represent the full width of the Universe, one could still be sure, but at the extreme distances being measured, the galaxies had grown so dim that it was almost impossible to work with them.

But then, in 1963, the Dutch-American astronomer Maarten Schmidt discovered the quasars, objects much brighter than galaxies and with spectral properties indicating them to be much farther than even the farthest known galaxy. Even the nearest quasar was of the order of a billion light-years away. A sphere centered on the Sun and large enough to include the nearest quasar would be two billion light-years in diameter at least:

11. 1963—2,000,000,000 light-years

This fivefold increase was not the end, for surely there would be more distant quasars. In 1973, in fact, the distance to one of them, known as OH471, was measured as twelve billion light-years. A sphere that centered on the Sun and included OH471 would represent a new record:

12. 1973—24,000,000,000 light-years

This is a further twelvefold increase.

In twelve stages, then, man's appreciation of the diameter of the Universe had risen from 5,000 miles to 24,000,000,000 light-years, an increase of nearly 30,000,000,000,000,000,000,000-fold in 2,500 years. This represents a doubling of the known size of the Universe every thirty-two years, on the average.

Of course, most of the increase came since telescopic times. Since 1671 the known size of the Universe has increased 80,000,000,000,-000 times in 302 years. This represents a doubling of the known size of the Universe over that period of time every 6.5 years on the average.

And we seem to be keeping it up. In the last ten years we have increased the known size of the Universe twelvefold, an amount rather above the average. So if we continue expanding the known size of the Universe at the rate we have been for the last three centuries, then by A.D. 2010 we ought to have driven the boundaries of the Universe outward and established the diameter of the known sphere in excess of the trillion-light-year mark.

Unfortunately, we won't.

After all the doubling and redoubling and re-redoubling, seemingly without end, astronomers have, indeed, reached the end, and as luck would have it, they reached that end in your lifetime and mine; indeed, in the good old decade of the 1970s.

How is that possible? Well, here goes, for I am now ready to talk about the expanding Universe and the receding galaxies.

In order to simplify the matter of the expanding Universe, why not reduce three dimensions to one. Everything remains valid and it is easier to visualize the argument in one dimension.

Let's begin by considering a string of lighted objects (micro-Suns if you like) stretching out in a straight line indefinitely to right and left. We must imagine that they're the only things in existence so that if any of the lights moves, you can relate that movement only to the remaining lights.

Let us next suppose that the lights are arranged at equal intervals and, for convenience's sake, let's call those intervals one mile. Let's imagine ourselves microbes attached to one of the lights, which we will call Light O (for both "zero" and "observer"), and that from that light we are capable of observing all the others.

To one side we see all the eastern lights and can measure their distances from us. The nearest one, one mile away, is E-1; the next one, two miles away, is E-2; the next one, three miles away, is E-3; and so on as high as you like—to E-1,000,000 or more, if you wish. (If the lights are in a straight line, then the first one blocks all the rest, of course, but we can pretend, for argument's sake, that they are all transparent and that we can concentrate on any one of them; ignoring those in front of or behind it.)

In the other direction, we have the western lights, and we can number and identify them in the same way: W-1, W-2, W-3, and so on, as high as you like.

We can define the positions in which the lights are placed by using small letters. Light E-1 is in position e-1; Light W-5 is in position w-5; and so on.

Now comes the crucial point. Let us suppose that in the course of some interval of time (for convenience's sake, let us say, in one second) the interval of space between each pair of neighboring lights doubles, and changes from one mile to two miles. In other words, the line of lights expands linearly.

Since only the lights exist, there is nothing to which to compare the motions of any light except the other lights. You, on your Light O, will have no sense of motion. You will feel motionless but you will see that E-1 has moved off to position e-2 and that W-1 has moved off to position w-2, each of them having receded from you at the not unbelievable speed of one mile per second.

This is precisely the situation all along the line of lights. An observer on *any* of the lights will see only a slow recession on the part of his immediate neighbors. Though the line is a sextillion miles long and there are a sextillion lights at one-mile intervals and though every single interval between all those lights has expanded from one mile to two miles in one second, an observer on every

single one of those lights would be conscious of only a slow recession
on the part of his immediate neighbors.

Of course, if an observer is standing somewhere else and could
see the entire sextillion-mile stream of lights as a whole and saw the
intervals all expand it would be plain to him that in one second
the length of the entire line had increased from one sextillion miles
to two sextillion miles and that some of those lights would therefore have had to move at many millions of times the speed of light.

However, there can't be an outside observer since we are assuming
that only the lights exist and that observers can only be on the
lights (or, at a pinch, anywhere on the straight line between the
lights). And even if an outside observer did exist, the rules of relativity would prevent him from seeing the entire stretch of line at
one time.

But suppose that while standing on Light O you observe, not
just the neighboring lights, but all the rest as well. We have assumed
this could be done.

Looking eastward from Light O, you see that E-1 has moved from
position e-1 to e-2. E-2, on the other hand, which is now separated
from you by two two-mile intervals instead of two one-mile intervals, has moved from e-2 to e-4. E-3, separated from you by three
two-mile intervals, has moved from e-3 to e-6; E-4 has moved from
e-4 to e-8; E-5 from e-5 to e-10; and so on indefinitely. Looking
westward, you see that W-1 has moved from w-1 to w-2; W-2 from
w-2 to w-4; and so on indefinitely.

Taking note of positions before and after, and knowing the time
interval in which that change has taken place, you decide that since
E-1 has moved from e-1 to e-2, it has receded from you at one mile
per second. Since E-2 has moved from e-2 to e-4, it has receded
from you at two miles per second. Since E-3 has moved from e-3 to
e-6, it has receded from you at three miles per second; and so on,
indefinitely. The same thing is happening to the western lights.

Because of the constant expansion of the line; the conversion of
every interval to one that is double its previous length, an observer
at Light O finds not only that every other light, in either direction,
is receding from him, but that *the rate of recession is proportional to
the distance from him.*

We can argue conversely. Suppose the observer knows nothing
about the expansion of the line. All he knows is that by measuring

the motion of the lights in either direction, he finds that all are receding from him and that the rate of recession is proportional to the distance from him. Having observed that, he must inevitably come to the conclusion that the line is expanding.

These same observations would be made, and these same conclusions would be arrived at, *no matter which light the observer was standing on.* Light O is not a unique light because all the others are receding from it. Other observers on any other light would find themselves in the same "unique" position.

Next, let us suppose that the speed of light is exactly 186,282 miles per second (omitting the extra 0.4 miles per second). We can say, then, that by the line of argument worked out just above, an observer on Light O would find E-186,282 (or W-186,282) to be receding from him at the speed of light; and that E-186,283 (or W-186,283) and all the lights beyond in either direction would be receding from him at speeds *above* the speed of light.

But how can this be? Doesn't Einstein say nothing can go faster than light?

We agree that light E-186,283 must recede from Light O at a speed greater than the speed of light, but that's *a calculated* speed worked out by logic. Can the speed actually be *measured?*

Suppose we are on Light O, observing all the other lights. We actually measure the speed of their recession by means of the red shift in their spectra. The light a receding object emits shows a red shift because there is a loss in energy in that light from the normal level in the light that the object would be emitting if it were motionless relative to you. The further the light and the more rapidly it is receding, the greater the red shift in the light it emits and the greater the energy loss.

Finally, by the time we observe E-186,282 (or W-186,282), the light it is emitting as it recedes from us at the speed of light shows an infinite red shift, a *total* loss of energy. There is no light to reach us. In other words, in the case of an expanding line we can only detect light and therefore only measure speeds of recession up to the point where an object is receding at the speed of light. Beyond that we cannot possibly see or measure anything. For in the line of lights we have postulated, E-186,282 and W-186,282 are the limits of the "observable Universe" for any observer on Light O.

We can never see any lights beyond that limit. We might speculate that, relative to ourselves, they are moving at more than the speed of light and that to any observer that can see them and *measure* their speeds of recession they would be moving at less than the speed of light.

In fact, from every light in the entire string, there is an observable Universe with limits slightly different from that which can be observed from every other light. (If we took the FitzGerald Contraction into account, we might argue that *all* the lights would be seen closer than the limit and that *none* would be beyond the limit—but the limit would remain.)

All of this, which I have worked out for a one-dimensional Universe of lights, works out also for the familiar three-dimensional Universe of galaxies in which we live.

The Universe is expanding at a constant rate. Each galaxy seems motionless to itself, and to each galaxy, the neighboring galaxies (or clusters of galaxies) seem to be receding at rates that are not too rapid. From each galaxy, the rate of recession of the other galaxies is seen to be increasing in direct proportion to the distance from the observer's galaxy. Furthermore, for each galaxy there is a limit marking off the observable Universe at that point where the galactic speed of recession is equal to the speed of light.

Suppose there are an infinite number of galaxies beyond that limit, all of them moving faster than light relative to ourselves. Those faster-than-light speeds cannot be measured and those faster-than-light galaxies cannot be detected.

The latest observations of galactic recessions make it look as though the rate of recession increases fifteen miles per second every million light-years of distance from us. That means that at a distance of 12,500 million light-years, the speed of recession is $12,500 \times 15$, or just about equal to the speed of light.

The radius of the observable Universe, then, is 12,500 million light-years and its diameter is 25,000 million light-years. Since we have now detected a quasar at almost this limit (which is receding from us at roughly 90 per cent the speed of light), we cannot expect to see further by more than a trivial amount. (That is why newspapers have spoken of astronomers having detected the end of the Universe.)

Unless—

Well, the Greeks came to a halt in 150 B.C. since they had probed the Universe as far as it was possible to go without a telescope (the name of the device meaning "to see the distant"). There is nothing mysterious about a telescope but it was inconceivable to the Greeks, and if we could put ourselves in their place, we might feel justified at supposing that any distance beyond that of the Moon was forever inaccessible to the human mind.

Can it be that we have now merely probed the Universe as far as it is possible to go without a "tachyscope" ("to see the very fast")? Perhaps there is nothing mysterious about a tachyscope, once it is devised, but right now it seems inconceivable to us. Right now, we seem to be justified in feeling that the distance of anything beyond Quasar OH471 is forever inaccessible to the human mind.

—But perhaps we're wrong, too.

3. Constant as the Northern Star

One trouble with making a profession of thinking of things is that occasionally you think of something that makes you recognize your own stupidity. This is particularly embarrassing when you spend half your time smiling modestly while other people tell you how smart you are.

Here is what I said in a book of mine called *The Land of Canaan* (Houghton Mifflin, 1971) concerning the early voyages of the Phoenicians, who were the most remarkable mariners of ancient times:

"It is possible that the Phoenicians were helped in their exploration by a notable advance in technology. The open sea lacks landmarks to guide the traveler for, as the word itself indicates, these are restricted to land. The stars in the sky might be landmarks if they did not steadily turn. They turn about a hub in the sky, however, and near that hub is the bright star Polaris, the one star that scarcely changes position at any time. It is possible that the Phoenicians were the first to learn to use Polaris as a 'landmark' at sea and that it was this that opened the western Mediterranean to them."

I should have known better, and it is small comfort to me that William Shakespeare made a similar mistake. In *Julius Caesar* he has Caesar make the following grandiloquent statement shortly before he was stabbed:

> But I am constant as the northern star,
> Of whose true-fix'd and resting quality
> There is no fellow in the firmament.
> The skies are painted with unnumber'd sparks;
> They are all fire and every one doth shine;
> But there's but one in all doth hold his place.

I think it is possible that Francis Bacon might not have made this particular mistake—another piece of evidence that Bacon didn't write Shakespeare's plays.* Well, let's see what the mistake is, and why.

The Earth rotates about an axis, which can be pictured as a line passing through the center of the planet and emerging at opposite points on the surface, at the North Pole and the South Pole. The west-to-east rotation of the Earth about this axis is unnoticeable to us. It seems to us that the Earth stands still and the sky rotates east to west.

The illusory rotation of the skies takes place around the same axis as does the real rotation of the Earth. If the axis of the Earth is extended in imagination it will intersect the sky, or the celestial sphere, at two opposite points, the North Celestial Pole and the South Celestial Pole. A line whose points are each and all halfway between the two celestial poles is the Celestial Equator and, indeed, the celestial sphere can be marked off with a cross-hatching of latitude and longitude just as the surface of the Earth is (in imagination only, of course).

Whenever we stand on Earth, the point in the sky directly over-head (at zenith) is at the same celestial latitude as the earthly latitude on which we stand.

If we were standing on the Earth's North Pole, for instance, we would see the North Celestial Pole at zenith and all the stars turning in a grand twenty-four-hour sweep about it in a counterclockwise direction. If we were standing on the Earth's South Pole, we would see the South Celestial Pole at zenith and all the stars turning a grand twenty-four-hour sweep about it in a clockwise direction.

From anywhere in the Northern Hemisphere, the North Celestial Pole is somewhere above the northern horizon. Where it is, it remains, and the stars circle it. The South Celestial Pole is some-where below the southern horizon and remains there. From anywhere in the Southern Hemisphere, it is the South Celestial Pole that is fixed in some point above the southern horizon, while the North Celestial Pole is hidden somewhere below the northern horizon.

* On the other hand, people might judge from the mistake about Polaris that Isaac Asimov, well known to be knowledgeable about astronomy, didn't write *The Land of Canaan*. Deductions of this sort can be shaky.

The nearer we are to either of Earth's poles, the higher the corresponding celestial pole is in the sky. The height of the celestial pole above the horizon is, in point of fact, equal to the observer's latitude on the Earth's surface.

At the moment, for instance, I am sitting at a point that is at roughly 40.8° north latitude. (The symbol ° is read "degrees.") This means that, for me, the North Celestial Pole is 40.8° above the northern horizon calculated for sea level.

All the stars circle the North Celestial Pole, and those that are less than 40.8° from it make circles that at no point reach the northern horizon. All those stars, as viewed from where I sit, remain always above the horizon, day and night.†

Stars that are more than 40.8° from the North Celestial Pole make circles so large that (as viewed by myself) they cut below the horizon. Such stars rise and set.

The South Celestial Pole, from where I sit, is always 40.8° below the southern horizon and any star within 40.8° of it makes a circle that remains below the horizon at all points. I never see any of them.

If I were to leave home and travel farther north, the North Celestial Pole would rise higher in the sky and the circular patch of stars that remained always above the horizon (and the similar patch of stars in the neighborhood of the South Celestial Pole that remain always below the horizon) would increase in area. At the North Pole, all the stars in the Northern Celestial Hemisphere would remain always above the horizon and all the stars in the Southern Celestial Hemisphere would remain always below the horizon. (The situation would be reversed at the South Pole, of course.)

If I were to leave home and travel farther south, on the other hand, the North Celestial Pole would sink lower in the sky and the patch of stars always above the horizon at one end and below the horizon at the other, would decrease in area. At the Equator those patches would shrink to zero. The North Celestial Pole would be at the northern horizon, the South Celestial Pole at the southern, and all the stars without exception would rise and set. —It was this

† The presence of the Sun is a complication since when it is in the sky, we can't see the stars. However, the ancients learned to calculate the positions of the stars even when they could not be seen.

changing behavior of the stars with changing latitude that helped give the ancient Greeks the notion that the Earth must be a sphere.

Naturally, anyone watching the skies attentively night after night from a reasonably high latitude‡ would see the stars make a circle about some point in the northern sky. That point, the center about which the stars turn, would itself be motionless, of course.

Unfortunately, there is no way in which a particular point on the featureless celestial sphere can be marked out easily, *unless* there should happen, by the sheerest of coincidences, to be a bright star at or near the point.

As it happens, there is: a star with a magnitude of 2.1, which makes it one of the two dozen or so brightest stars in the sky. Because this star is so near the North Celestial Pole, it is called "Polaris" or, less formally, the "Pole Star." Because the star remains always in the near neighborhood of the North Celestial Pole, which is itself fixed in the northern part of the sky, the star is called, even more informally, the "North Star."

Polaris is not exactly *at* the North Celestial Pole, however. It is about 1.0° away. This means it makes a small circle about the North Celestial Pole that is 2.0° in diameter. Such a circle is four times as wide as the full Moon, so it is by no means an insignificant motion.

However, a change in position becomes noticeable only when it can be compared to something that doesn't change. Since all the stars turn with the sky in one piece, Polaris remains in fixed relationship to the other stars and they won't serve as comparison. The one convenient reference point unaffected by the motions in the sky is the horizon. The height of Polaris above the northern horizon varies in the course of a twenty-four-hour period (from where I sit) from 41.8° to 39.8°.

This difference is not large enough to impress the casual observer. Nor does the fact that Polaris, during this same period, varies from 1.0° east of the true north to 1.0° west prove impressive. To the casual observer, the swing goes unnoticed and it appears that Polaris remains fixed in position, marking out always the exact north.

‡ The first careful star-observers were probably the Sumerians, who lived at about 32° N., far enough northward to make the North Celestial Pole a dominant factor in the night sky.

That is why you can speak of something being as "constant as the northern star" and can say that of all the stars "but one in all doth hold his place." And that is why you can always tell which direction is north (provided the night is clear) and can steer a ship surely when out of sight of land—as I said the Phoenicians could.

So where's the mistake? Well, let's go on—

In addition to rotating on its axis, the Earth revolves about the Sun. To us on Earth, it seems that the planet remains motionless and that the Sun slowly changes its position among the stars. Careful observers, calculating the position of the stars (invisible in the neighborhood of the Sun, of course) will note that the Sun follows a huge circle all around the celestial sphere, taking 365¼ days to complete a single turn. The path it follows is called the "ecliptic."

As it happens, the ecliptic does not follow the Celestial Equator as one might assume. Instead it cuts across the Celestial Equator at two points on opposite sides of the celestial sphere. The ecliptic cuts across the Celestial Equator at an angle of 23.5°.

For half the year the Sun follows the line of the ecliptic in a loop north of the Celestial Equator. It is then spring and summer in the Northern Hemisphere and the days are longer than the nights. The other half the year, the Sun follows the ecliptic in a loop south of the Celestial Equator and it is then fall and winter in the Northern Hemisphere with the nights longer than the days. (The situation is reversed in the Southern Hemisphere.)

The two points where the ecliptic crosses the Celestial Equator are the equinoxes ("equal nights," because when the Sun is at those points, day and night are equal in length the world over). The particular crossing point at which the Sun is moving from south of the Celestial Equator to north of it, is the vernal equinox because that marks the beginning of spring in the Northern Hemisphere— where the ancient astronomers lived. The other, in which the Sun is moving from north to south, is the autumnal equinox since this marks the beginning of fall.

In passing from vernal equinox to autumnal equinox, the Sun follows the line of the ecliptic in its northern loop and halfway between the equinoxes it reaches its most northern point, the summer solstice ("Sun-stationary," because at that point the Sun is momentarily stationary in its northward drift before reversing to a southward drift). Between the autumnal equinox and vernal equinox, the Sun follows the southward loop of the ecliptic, reaching the southernmost point, the winter solstice halfway between the equinoxes.

In terms of man-made dates, we have the vernal equinox on March 20; the summer solstice on June 21; the autumnal equinox on September 23; and the winter solstice on December 21.

All this is explained by the fact that the Earth's axis of rotation is not exactly perpendicular to the direction of the Sun, but is tipped 23.5° to that perpendicular. The direction of the axis remains constant relative to the stars as the Earth revolves about the Sun, the North and South Poles pointing to the same spots in the celestial sphere throughout the entire revolution.

On June 21, at the summer solstice, the direction of the axis is such that the North Pole is tipped 23.5° in the direction of the Sun (and the South Pole therefore 23.5° in the direction away from the Sun). On December 21, when the Earth is at the opposite end of its orbit about the Sun, the North Pole, tipped in the same direction relative to the stars, is now tipped away from the Sun, while the South Pole is tipped toward it.

The fact that the Earth's axis steadily changes its direction relative to the Sun would make no difference to the manner in which the Sun attracts the Earth, if the Earth were a perfect sphere.

But the Earth *isn't* a perfect sphere. Because the Earth rotates, there is a centrifugal effect that tends to counter gravity and lift the surface. The effect is more pronounced the faster the Earth turns. The parts of the Earth farther from the axis turn faster, and there is more and more of Earth farther and farther from the axis as one goes from the Poles to the Equator.

The centrifugal effect is therefore stronger and stronger as one goes toward the Equator, and the substance of the Earth lifts up against gravity more and more. The surface of the Earth is farther from the center the closer one gets to the Equator, and at the Equator the surface is thirteen miles farther from the center of the Earth than is true of the surface at the Poles. Since the bulge is most extreme at the Equator, it is usually called an "equatorial bulge."

This means that the Earth, as seen from the Sun, is *not* symmetrical. At the summer solstice, the North Pole is tipped toward the Sun, and the equatorial bulge does not lie across the center of the cross-sectional circle of Earth as seen from the Sun. It curves south of the center and the Sun's gravitational pull tends to lift it northward. Of course, there is the equatorial bulge on the other side of the Earth which curves north of the Equator so that the Sun's gravity tends to pull it southward. That, however, does not restore the symmetry, for the other side of the Earth is 8,000 miles farther from the Sun, and the Sun's gravitational pull is a trifle weaker there because of the extra distance. As a result, the Sun pulls the near side northward a little harder than it pulls the far side southward, and there is thus a net force northward.

When the Earth is on the other side of the Sun, the net force is southward. During half the year, then, the net force on the equatorial bulge is northward, increasing from zero at the vernal equinox to a maximum at the summer solstice and back to zero at the autumnal equinox. During the other half of the year, the net force is southward, increasing from zero to a maximum at the winter solstice and decreasing to zero again.

The Sun is not the only body in the solar system that exerts such an unbalanced pull on the Earth. The Moon does it, too. The path of the Moon among the stars, as seen from Earth, is closer to the ecliptic than it is to the Celestial Equator. (It is tipped to the ecliptic by five degrees. When the Moon and Sun are both near one of the two crossing points, the Moon crosses in front of the Sun and there is an eclipse. It is because the Moon's approach to the ecliptic is required for an eclipse that the ecliptic is so called.)

The Moon also yanks at the equatorial bulge. To be sure the Moon's mass is only 1/27,000,000 that of the Sun, and even allowing for the Moon's much smaller distance from Earth, its gravitational pull on Earth is only a little over 1/200 that of the distant Sun.

However, the unbalanced pull on the equatorial bulge is related to the *difference* in gravitational pull on the two sides of the Earth, and that varies more strongly with distance than the over-all pull itself does. As a result, the Moon's contribution to the asymmetric yank at the equatorial bulge is 2.2 times that of the Sun.

This lunar-solar yank at the equatorial bulge causes the Earth's axis to wobble in just the way a gyroscope would under similar circumstances. (Indeed, the Earth, like any massive, rapidly rotating object, *is* a gyroscope.) Without trying to explain why, here's what happens—

The axis of rotation remains tipped 23.5° to the perpendicular, but moves in such a way that the North Pole moves in a circle around the perpendicular. (So does the South Pole, of course.)

Another way of describing it is that the point of the axis at the center of the Earth remains motionless, but north and south it marks out a cone.

Let's look at the sky and see what this means. If the Earth's axis were directly perpendicular to the direction of the Sun, it would point at a spot on the celestial sphere which would be equidistant from every point on the ecliptic. That spot would be to the ecliptic as Earth's North Pole is to its Equator. Hence the spot in the sky to which the axis would be pointing if it were perpendicular to the direction of the Sun is called the "North Pole of the ecliptic." (On the opposite side of the celestial sphere is the "South Pole of the ecliptic.")

The North Celestial Pole is 23.5° from the North Pole of the ecliptic, but because of the unbalanced pull of the Sun and the Moon on the equatorial bulge, it describes a circle about the North Pole of the ecliptic, remaining always 23.5° away as it does so.

As the North Celestial Pole makes this circle, and as the South Celestial Pole makes a similar circle at the other end of the sky, the Celestial Equator shifts position, too. It does so in such a way that as the celestial poles make one complete circle, each crossing point of the Celestial Equator and ecliptic (each equinox, in other words) makes a complete circle about the sky, too.

Each crossing point moves slowly from east to west as the axis makes its wobble. Meanwhile the Sun is moving from west to east toward the equinox. Since the crossing point is coming to meet it, the Sun reaches the equinox earlier than it would if the crossing point weren't moving. The actual equinox precedes (very slightly) the ideal one.

This was first discovered by the Greek astronomer Hipparchus about 130 B.C., and he called it the "precession of the equinoxes." The cause of this precession wasn't explained till Isaac Newton did the job eighteen centuries later.

The precession of the equinoxes is very slow and it takes the position of each crossing point 25,780 years to run a complete circle about the ecliptic. This is the same as saying that the North Celestial Pole takes 25,780 years to complete its circle about the North Pole of the ecliptic (and the South Celestial Pole about the South Pole of the ecliptic).

There are 360° in a circle, so the North Celestial Pole moves about 0.014° per year. That isn't much. It would take the North Celestial Pole thirty-seven years to move the width of the full Moon, and this motion would certainly not be noticed under merely casual observation. It also means that the vernal equinox comes each year twenty minutes earlier than it would have done had there been no precession.

Since the North Celestial Pole *does* move, however, it affects the position of Polaris (which also moves relative to the other stars against the sky, but far more slowly even than precession). At the present moment, the North Celestial Pole is moving in the direction of Polaris, so that Polaris is making smaller and smaller circles about the North Celestial Pole and is becoming more and more nearly a true North Star.

At just about A.D. 2100 the North Celestial Pole will skim past Polaris, and at that point Polaris will be only 0.47° away and will make a circle with a diameter about twice that of the full Moon. After that, though, the North Celestial Pole will be moving away from Polaris and the circles it makes will increase in diameter again.

Since the North Celestial Pole has been steadily approaching Polaris for centuries, it must have been further away in the past. In 1900 Polaris was 1.2° from the North Celestial Pole, instead of the 1.0° distance it is today.

Suppose we go further back, to 1599, when Shakespeare was writing *Julius Caesar*. At that time Polaris was 2.9° from the North Celestial Pole. The circle it made every twenty-four hours was 5.8° wide, about nine times the width of the full Moon. From the latitude of London (51.5°), Polaris' position varied from a height of 54.4° above the northern horizon to 48.6° above.

A casual observer might still not notice this. Nor might he notice that in the course of a day Polaris wobbled east and west to extremes of 2.9° east and west of the true north. Shakespeare could still speak of something being as "constant as the northern star" and get away with it.

But hold on. Shakespeare was having Julius Caesar say it just before his assassination on March 15, 44 B.C., and where was Polaris then?

I'll tell you where it was. It was 12.2° from the North Celestial Pole. *Twelve point two degrees!*

From the latitude of Rome (42.0°) Polaris would, in 44 B.C., have been seen to make a huge circle 24.4° in diameter. In the space of twelve hours it would have dropped from a height of 54.2° above the northern horizon to a height of only 29.8° above it. With its height above the horizon cut in half, no one could possibly consider it as "constant."

But wait. Just because Polaris wasn't near the North Celestial Pole doesn't mean that some other star might not be and that this other star might not have been the one referred to by Julius Caesar. Well, there wasn't. No star bright enough to be more than just barely seen was nearer to the North Celestial Pole than Polaris was in 44 B.C.

What it amounts to then was that in Julius Caesar's time, *there wasn't any North Star!* And even if the real Caesar spoke as grandiloquently as the stage Caesars did in Shakespeare's time (which he most certainly didn't), he wouldn't have compared his constancy to something that didn't exist.

And since the *fact* of the precession of the equinoxes had been known for over seventeen centuries, even if the explanation was still eighty years off, Francis Bacon, a very learned man, might have avoided the mistake.

And for myself? What about the Phoenicians?

The great age of their navigation began about 1000 B.C. and at that time, Polaris was 16.8° from the North Celestial Pole. Of those stars which, in 1000 B.C., were closer to the North Celestial Pole than Polaris was, Thuban is the brightest. Thuban is the brightest star of the constellation of Draco (the Dragon) which occupies the region of the North Pole of the ecliptic. But Thuban has a magnitude of 3.6 so that it is only one fourth as bright as Polaris. What's more, it was still fully 9° from the North Celestial Pole, so that it wasn't much of a North Star either.

The Phoenicians could not, therefore, have guided their navigation by discovering the stationariness of Polaris or of any star, and my careless statement in *The Land of Canaan* is wrong.

But, then, what was it the Phoenicians *did* discover? Anything at all?

Yes, they probably made one important discovery—in the northern sky, there are seven second-magnitude stars that form a distinct and familiar shape we call the Big Dipper. All of it lies between 20° and 40° of the North Celestial Pole, so that all of it remains in the northern sky constantly, never setting, as seen from Europe or from the northern United States.

It is far easier to see and locate a collection of seven stars in a distinctive shape than it would be to see and locate any single star. Any person, if he can see the stars at all, can locate the Big Dipper at once and without trouble, though he might not be able to locate Polaris (except by using the Big Dipper as a reference).

Furthermore, because of the distinctive shape of the Big Dipper, it is easy to see that it is turning in the sky, because the dipper shape is now right side up, now on end, now upside down and so on. It is easy to imagine the center about which the Big Dipper is turning and use that as a rough north point.

What's more, the North Celestial Pole, as it has been moving along in the course of the precession of the equinoxes, has been skimming along the edge of the Big Dipper at a practically constant distance for something like six thousand years. It is only in the last thousand years, in fact, that the North Celestial Pole is beginning to pull away from the Big Dipper.

It follows, then, that over all of civilized history, no matter whether there was a North Star or not, the Big Dipper has been spectacularly circling the sky and marking out the north.

The Greeks saw the Big Dipper as a wagon, or "wain," and also saw it as part of a bigger stellar configuration they imagined to be in the shape of a bear. In the time of the Homeric poems, for instance, there was no mention of the North Star, for there wasn't one, but there was mention of the Bear. Thus, in the fifth book of the *Odyssey*, when Odysseus is leaving the island of Calypso to return home, Homer says:

"He never closed his eyes, but kept them fixed on the Pleiades, on late-setting Boötes, and on the Bear—which men also call the

Wain, and which turns round and round where it is, facing Orion, and alone never dipping into the stream of Oceanus, for Calypso had told him to keep this to his left."

If Odysseus kept the Big Dipper to his left, he would be heading toward the east, which was where he wanted to go. If he kept it to his right, he would go west. If he headed toward it, he would go north, and away from it, south. But the *Odyssey* was written about 800 B.C. at the very earliest, and this use of the Big Dipper was probably what the Phoenicians discovered about 1000 B.C.

—So why didn't I take a few moments to think this through when I wrote that passage in *The Land of Canaan?*

4. Signs of the Times

I've just come back from the University of Delaware, where I gave a talk on the significance of science fiction. What with dinner, followed by an interview by newsmen, followed by the talk, followed by a reception, I spent five hours with people, talking all the time.

I can only be glad that I don't mind talking all the time.

The reception which followed the talk and which lasted two hours, consisted of a question-and-answer period, with no holds barred and with everything completely informal. Naturally, answering off the top of my head, I sometimes get my foot firmly implanted in my mouth. Not often, thank goodness, but this time I turned out a whopper.

I was asked if I enjoyed giving talks and, utterly forgetting tact, I said, "I love to give talks but I love to write even more and it is only with extreme reluctance that I quit my typewriter to visit campuses. You have no idea how difficult it was to persuade me to come here."

And the silence that followed was quickly broken by one student who responded austerely, "It was difficult to persuade us, too."

It served me right, of course, and I could only join in the laughter at my expense, with a face which (I hope) wasn't quite as red as it felt—but probably was.

The incident made me think about the Gentle Readers of these, my humble essays. I write these essays, primarily, because they amuse me, but now I wonder if, on occasion, it may not be rather difficult to persuade you to accompany me (as the student at the University of Delaware implied).

The previous chapter, on the face of the celestial sphere, done as it was without diagrams, may, for instance, have been hard to swallow.

If so, please forgive me in your Gentle way, for I haven't finished. In this chapter I want to continue with the effect of the precession of the equinoxes on some details of the celestial sphere and on that tissue of absurdity, astrology.

To begin with, let's turn back to the ecliptic, which I mentioned in the previous chapter as marking out the apparent yearly path of the Sun against the starry background of the celestial sphere.

To make it easier to consider that background, those stars which can be seen from the north temperate zone have been grouped into patterns called "constellations" by the ancient stargazers. The constellations we now recognize are essentially those used by ancient Greek astronomers.

The constellations do not have real existence, of course, for the stars that make them up have no interconnection, by and large, but are strewn helter-skelter over the surrounding hundreds of light-years. The configurations happen to be what they are only because we are looking at the sky from a certain place and, since the stars (including our own Sun) are all moving, at a certain time. Shift our position a thousand light-years in space or a million years in time and the sky would be unrecognizable. The Greek astronomers, however, assumed the constellations to have real existence—made up of eternally fixed points of light attached to a solid firmament. Modern astrologers, who retain a distorted-Greek astronomy, act as though they believe the same (and maybe some really do).

The path of the ecliptic passes through twelve of these constellations, so that the Sun remains in each for roughly a month. In fact, the division was probably deliberately set at twelve for this purpose since the month was the chief unit of time in the lunar calendars used by the ancient Babylonian and Greek stargazers.

This means that the Moon makes one circle of the sky while the Sun passes through a single one of these constellations. (This is only approximately true, but close enough to satisfy primitive astronomers and modern astrologers.) Besides 12 is an easy number to divide evenly by 2, 3, 4, and 6—an important consideration for those without an efficient system of number symbols, such as the ancient Babylonians and Greeks.

The names of the twelve constellations are in Latin even today but all have common English translations. In the order in which the Sun passes through them they are: (1) Aries, the Ram; (2) Taurus, the Bull; (3) Gemini, the Twins; (4) Cancer, the Crab; (5) Leo, the Lion; (6) Virgo, the Virgin; (7) Libra, the Scales; (8) Scorpio, the Scorpion; (9) Sagittarius, the Archer; (10) Capricornus, the Goat; (11) Aquarius, the Water Carrier; and (12) Pisces, the Fishes.

Because seven of the twelve constellations are imagined in the figures of animals (eleven if you count human beings as animals, leaving only Libra as inanimate), they are referred to, all together, as the "zodiac," from Greek words meaning "circle of animals."

The star configurations don't really resemble the objects they are named for. It took a most lively and metaphoric imagination to see them, but I suppose the less sophisticated Greeks thought that pictures of rams and bulls, and perhaps even the real things, existed in the sky. It may be that modern astrological devotees think so, too, assuming they think at all.

The ancients, in constructing the constellations, made no attempt to have them take up fixed and equal fractions of the celestial sphere. They grouped them into what seemed natural star-combinations so that some constellations are large and sprawling and others are quite compact. Virgo, for instance, covers much more space in the sky than Aries does.

What's more, the Sun, in making its way along the zodiac, crosses some constellations along a wide diagonal, others along a relatively narrow corner. The Sun, therefore, does *not* remain equal times in each constellation.

Modern astronomers have fixed the boundaries of the constellations on the celestial sphere (including those constellations near the South Celestial Pole which were only observed by Europeans in modern times), following as best they could the groupings as described by the ancients. These boundaries, convenient as reference points in astronomy, are now universally adopted by astronomers, and if we follow those we can work out how long the Sun remains within each constellation of the zodiac (see Table 1).

As you see, the Sun is in Virgo for almost seven weeks, while it is in Cancer for only three weeks. Scorpio is the queerest case. In the interval between Libra and Sagittarius, the Sun is in Scorpio for only six days! For eighteen days thereafter, if we go by the established boundaries of the constellations, the Sun is in Ophiuchus, the Serpent Bearer, which is not considered a constellation of the zodiac at all by the astrologers.

None of this fine detail of constellation inequality is, of course, given any attention whatever by astrologers. It may be that to do so would place undue strain on their mathematical resources. Less cynically, it might be reasoned that astronomical boundaries of the constellations are merely man-made and need not be given credence. This is true, of course, but so also are the constellations themselves purely man-made, as is the convention that divides the ecliptic into twelve parts, rather than ten or one hundred.

In any case, astrologers make it easier for themselves by pretending that the constellations are equal in width and that the Sun remains an equal number of days in each. That simplifies the mathematics and reduces the strain on the astrologer.

In order to account for the fact that when astrologers speak of the "Sun in Aries," it may really not be in Aries, as might be pointed out by some mocking astronomer, there is an astrological convention that wipes out the constellations altogether. The astrologers speak of the *signs* of the zodiac. These signs have the same names as the constellations but have no connection with them. The twelve signs of the zodiac are all equal in size and the Sun remains an equal length of time in each. It then doesn't matter whether the Sun is in the constellation of Aries or not; the astrologer says it is in the *sign* of Aries.

TABLE 1

Constellation	Passage of Sun (days)
Aries	22
Taurus	35
Gemini	26
Cancer	21
Leo	38
Virgo	47
Libra	25
Scorpio	24*
Sagittarius	34
Capricorn	30
Aquarius	24
Pisces	39

* For 18 of these days it is actually in Ophiuchus.

That accounts for the fact that people of every degree of ignorance and mis-education go around eagerly asking each other, "What's your sign?" and receiving as an answer the name of a constellation.

In this way, on the basis of the imaginary constellations, then, astrologers have built up a still more imaginary system of signs with which to impress fools and out of which to make a buck.

The ecliptic itself remains nearly fixed over the eons since it is a reflection of the plane of revolution of the Earth about the Sun and this doesn't change much. (The Greeks, of course, believed the Sun *really* moved along the ecliptic and I wouldn't be surprised if some astrologers believed that, too.)

The position of the Sun affects the seasons and the lengths of day and night, in accordance with the relationship of the ecliptic to the Celestial Equator, and the position of the Celestial Equator shifts with the precession of the equinoxes (which I discussed in the previous chapter).

The points where the Celestial Equator crosses the ecliptic are the equinoxes ("equal nights," because at that time, day and night are equal in length.) The Sun is at one of those points on March 20 and at the other, six months later, on September 23.

If we concentrate on those equinoxes, we find that their positions relative to the stars slowly shift as the Earth's axis wobbles (hence "precession of the equinoxes"). In a period of 25,785 years the equinoxes moves completely around the ecliptic, moving from east to west in the direction opposite to that in which the Sun moves along the ecliptic.

The length of time during which either equinox remains within a particular constellation of the zodiac depends upon the width of that constellation along the line of the ecliptic and is easily calculated (see Table 2). Of course, if we want to even out the widths of the constellations, we can say that an equinox remains within any given constellation of the zodiac for 2,148 years.

Let's consider the equinox that comes on March 20. This is usually referred to as the "vernal equinox" because it marks the beginning of spring by the conventions of the north temperate zone. (It marks the beginning of autumn in the south temperate zone, but we northerners have them southerners outnumbered.)

At the present moment, when the Sun marks the vernal equinox by crossing the Celestial Equator on its way northward, it is in the constellation Pisces, somewhat west of the center. Each successive vernal equinox, the point of crossing moves to 0.014° (or 0.84 minutes of arc) farther west. Eventually, some time in the future, it will slip into Aquarius; and if we look backward into the past, it was once in Aries.

In fact, if we accept the now-conventional boundaries of the constellation, the point of the vernal equinox was located exactly at the western boundary of Aries at about 100 B.C. and had been in Aries, progressively farther eastward, for fifteen hundred years previously, during all the time that astrological speculations had developed and grown more sophisticated in Babylonia and Greece. Since the vernal equinox is one logical place at which to begin the year (though we Westerners now use another), it became customary to start the list of constellations of the zodiac with Aries. Astrologers still do, though the excuse is now two thousand years out of date.

TABLE 2

Constellation	Passage of equinox (years)
Aries	1,550
Taurus	2,470
Gemini	1,840
Cancer	1,480
Leo	2,680
Virgo	3,320
Libra	1,760
Scorpio	1,700*
Sagittarius	2,400
Capricorn	2,125
Aquarius	1,700
Pisces	2,760

If we concentrate on the situation as it was in 100 B.C., we can say that the Sun entered Aries at the moment of the vernal equinox,

* For 1,225 years of this period, the equinox is actually in Ophiuchus.

passed eastward through Aries' full width, then went on through Taurus, Gemini, and so on.

Since the Sun passes through Aries in twenty-two days, it remains in that constellation from March 20 to April 11, at which time it enters Taurus, where it remains for thirty-five days, and so on. Of course, if we even out the widths of the constellations and use the sign instead, the Sun enters the *sign* of Aries on March 20 and then stays in each *sign* for just one twelfth of a year, or not quite 30.5 days. In Table 3 you will find the day on which the Sun enters each constellation and each sign of the zodiac—in 100 B.C.

As far as the constellations are concerned, the situation described in Table 3 is characteristic only of the decades in the immediate neighborhood of 100 B.C. The Sun enters Aries progressively earlier in the year in the period before 100 B.C. and progressively later in the year in the period since, thanks to the precession of the equinoxes, but astrologers, having established the *signs* of the zodiac as of 100 B.C., have never changed them.

To this very day, to this moment at which I am writing, the Sun is considered to enter the sign of Aries at the time of the vernal equinox. If you will look in your daily paper for the almost inevitable astrology column, you will find the days allotted to each

TABLE 3

Constellation or sign	Sun enters sign in 100 B.C.	Sun enters constellation in 100 B.C.
Aries	March 20	March 20
Taurus	April 20	April 12
Gemini	May 22	May 17
Cancer	June 22	June 12
Leo	July 23	July 3
Virgo	August 22	August 10
Libra	September 22	September 26
Scorpio	October 22	October 21
Sagittarius	November 22	November 14
Capricorn	December 22	December 18
Aquarius	January 22	January 17
Pisces	February 21	February 10

sign to be those given in Table 3 (give or take a day here and there).

We have seen that the actual position of the Sun in the zodiac shifts steadily relative to our calendar as a result of the precession of the equinoxes. Every 70.6 years, the zodiacal position of the Sun moves in such a way as to move the equinox forward one day. Thus, by 29 B.C. the Sun was in the actual constellation of Aries from March 21 to April 12; by A.D. 41 it was in the constellation of Aries from March 22 to April 13; and so on.

At the present moment, the position of the Sun has shifted forward twenty-nine days since 100 B.C., so that as of this year (1973), the Sun is in the constellation of Aries from April 18 to May 10. The time at which the Sun enters the various constellations *right now* is shown in Table 4.

Despite the twenty-nine-day shift, however, the astrological signs of the zodiac remain fixed and unaffected by the precession of the equinoxes, remaining as they were in 100 B.C.

In 100 B.C., when the two, constellations and signs, were most nearly in agreement, the Sun was in the same constellation and sign on 277 days of the year. By now, constellation and sign agree on only 106 days of the year. That there are as many as 106 is true only because of the unequal widths of the constellations. If we allowed the Sun to enter the constellation of Aries on April 18, as it does these days, and then considered all the constellations to be of equal width, then on only twenty-four days of the year would the sign and the constellation match and in a hundred forty years or so, sign and constellation would never match.

You might wonder why astrologers don't take the precession of the equinoxes into account. The reason cannot really be the fear of mathematical complication since the matter can be handled by any bright fourteen-year-old and, therefore, by *some* astrologers. It must be laziness.

And it gives away the folly of astrology. If the position of the Sun among the constellations has significance at all, then surely it is the actual position *now* that counts, not the position as it used to be in 100 B.C. If, on the other hand, the actual position doesn't matter, why should any other position?

Let's continue to focus on the vernal equinox. In 100 B.C., as I said, the Sun reached the vernal equinox at the western edge of Aries and traveled through the full width of the constellation before

TABLE 4

Constellation or sign	Sun enters sign	Sun enters constellation now
Aries	March 20	April 18
Taurus	April 20	May 11
Gemini	May 22	June 15
Cancer	June 22	July 11
Leo	July 23	August 1
Virgo	August 22	September 8
Libra	September 22	October 15
Scorpio	October 22	November 19
Sagittarius	November 22	December 14
Capricorn	December 22	January 16
Aquarius	January 22	February 15
Pisces	February 21	March 10

reaching Taurus. Since then, however, the Sun has been reaching the vernal equinox at a point on the Celestial Equator farther and farther west into the constellation of Pisces.

Of course, the fact that the vernal equinox passed into the constellation of Pisces in 100 B.C. is a matter of the actual boundary between Aries and Pisces as arbitrarily determined by modern astronomers. The boundary was vaguer in ancient times and one can imagine it to have existed 1.4° further west without any trouble. In that case, the point of the vernal equinox would have moved into Pisces in, say, 4 B.C.

Does that matter?

To an astrologer, it certainly does.

One can write, in Latin letters, the Greek phrase *Iesous Christos, Theou Uios, Soter*, which means, in English, "Jesus Christ, Son of God, Savior." The five Greek words begin, respectively, with the Greek letters: iota, chi, theta, upsilon, sigma. Stick those five letters together and they spell (in Latin letters) *ichthus*, which is Greek for "fish." For that reason, the early Christians used a fish as a symbol for Jesus, when more open avowal of their faith might have been dangerous.

Well, then, isn't it interesting that in 4 B.C., at the traditional year of the birth of Jesus, the vernal equinox moved into the constellation of Pisces, the *Fish?* Surely, anyone who thinks that one of God's major tasks in creating the Universe was to arrange the stars for the purpose of spelling out childish cryptograms would have to be impressed.

But let's leave the Aries-Pisces boundary at 100 B.C. (that wouldn't bother mystics who find a hundred-year discrepancy a mere bagatelle in any case) and calculate the times at which the vernal equinox reached other boundaries and passed into other constellations. We can do that by either taking the actual widths of the constellations as agreed on by modern astronomers or by pretending the constellations are all of equal width (see Table 5).

Of course, if we are going to have the vernal equinox, as it enters Pisces, signify the birth of Jesus, we have every right to suppose that at every new constellation-entry something equally significant in man's history is indicated. (Why else should immense stars, spread out at nine light-year intervals over many thousands of cubic light-years, be created except to spell out things obscurely to slowly developing primates on our minute planet?)

Thus, if we use equal-width constellations, the vernal equinox entered Taurus, the Bull, in 4395 B.C. Perhaps that was the time at which bull worship began in ancient Crete. I don't know that it was or that the doings in a little island merited the attention of the entire sky, but who am I to argue with Taurus, the Bull?

Then in 2247 B.C. the vernal equinox entered Aries, the Ram. Since that is only three centuries earlier than the time of Abraham, according to present interpretations of the biblical legends and since Abraham sacrificed a ram instead of his son Isaac—surely this must have something to do with the origin of Judaism. Who can doubt it?

The entry into Pisces I have already discussed. And now in A.D. 2049, only about three fourths of a century into the future, the vernal equinox will pass into Aquarius (if we calculate on an equal-constellation-width basis—on an actual-constellation-width basis, it won't happen for some seven centuries from now).

Well, what will happen in A.D. 2049? Aquarius, the Water Carrier, is usually represented as a man pouring water out of a vase, and this may symbolize the fact that the heavens will pour peace and plenty

TABLE 5

Vernal equinox enters

Constellation	actual constellation	equal-width constellation
Taurus	4410 B.C.	4395 B.C.
Aries	2570	2247
Pisces	100	100
Aquarius	A.D. 2660	A.D. 2049
Capricorn	4360	4197
Sagittarius	6485	6345
Scorpio	8885	8493
Libra	10585	10641
Virgo	12345	12789
Leo	15665	14937
Cancer	18345	17085
Gemini	19825	19233

upon the Earth. I strongly suspect that this is the origin of the idiot-song about the coming of the "Age of Aquarius," though I would cheerfully give odds of ten thousand to one that any particular person singing it hasn't the faintest notion of why it is called the "Age of Aquarius," if I were a betting man.

Of course, the great advantage of astrology is that it can never be shown to be wrong. If the next century destroys us, those who survive will point out that Aquarius symbolized the rain of radioactive fallout from the heavens and everyone will marvel at how well astrology works.

Everything I say about the vernal equinox holds for the autumnal equinox (which is the vernal equinox of the Southern Hemisphere) except that you have to shift the constellations, or the signs, by six.

In the days when the point of the vernal equinox was to be found at the western edge of Aries, the Ram, the point of the autumnal equinox was found (assuming constellations of equal size) at the western edge of Libra, the Scales. Of course, it isn't there anymore. It is now in the constellation of Virgo.

Halfway between the equinoxes are the solstices. At these points, the motion of the Sun away from the Celestial Equator ceases and there is a momentary period of motionlessness before it begins to drift toward the Celestial Equator again. It is at that stationary point, where the Sun reaches its maximum northerliness or southerliness that we place the solstice, which is from Latin words for "stationary Sun."

The solstice at which the Sun reaches its most northerly point comes on June 21. In the Northern Hemisphere, the day is at its longest, the night at its shortest, and the summer begins. To us of the north, then, this is the summer solstice.

At the other solstice, which the Sun reaches on December 21, the Sun is in its most southerly position and, in the Northern Hemisphere, the day is shortest, the night longest, and winter begins. So this is the winter solstice. (Need I tell you that the situation is reversed in the Southern Hemisphere?)

The summer solstice comes just three months after the vernal equinox and the Sun has a chance to pass through three constellations of the zodiac in that interval. In 100 B.C. the Sun, starting at the western edge of Aries, passes through the constellations of Aries, Taurus, and Gemini. Assuming the constellations to be of equal width, the Sun enters the constellation of Cancer, the Crab, at the moment of the summer solstice. (For two thousand years before that the point of the solstice was in the interior of Cancer.)

At the summer solstice the Sun is 23.5° north of the Celestial Equator and shines directly down upon those points of Earth which are at 23.5° north latitude. That parallel of latitude is called a "tropic" from a Greek work meaning "to turn" because when the Sun moves that far northward on the celestial sphere, it has gone as far as it can, turns, and begins to move southward. And because it makes its turn just as it enters the constellation of Cancer (at least in 100 B.C.) the line of 23.5° north latitude is called the "Tropic of Cancer."

If we start with Libra, which the Sun enters at the autumnal equinox and count three constellations, we see that it passes through Libra, Scorpio, and Sagittarius and, just as the winter solstice is reached, enters Capricorn. So the line of 23.5° south latitude is called the "Tropic of Capricorn."

Between these two imaginary lines on our globe lies the "Tropic Zone." And if you look at any geography book of Western origin written since the days of the Greeks, you will find those two tropics are presented with the names they had been given in the old Greek days—Cancer and Capricorn.

By now, though, you don't need me to tell you that those are misnomers today. When the Sun is shining directly over the Tropic of Cancer, it is in the constellation of Gemini; and when it is shining directly over the Tropic of Capricorn, it is in the constellation of Sagittarius.

Geography, like astrology, but with far better reason (does it matter whether you call it Tropic of Cancer or Tropic of Gemini?), does not recognize the precession of the equinoxes.

Well, I suppose it doesn't matter in astrology either. After all, do you really think that the position of the Sun against the stars affects you one way or the other? If so, is it the position of the Sun *now* that matters or the position that it had some two thousand years ago?

My feeling is that the rational content of astrology is zero if the present position of the Sun against the stars is considered and is not one whit more if the two-thousand-year-past position of the Sun is considered instead.

B—SOLAR SYSTEM

5. The Eclipse and I

Most of the knowledge of the Universe around me, in both its animate and inanimate aspects, I gain by hearsay, since my actual life is spent in very much an ivory-tower fashion. The result is that I constantly surprise myself with the extent to which I don't really accept emotionally what I know intellectually.

I was made keenly aware of this for the first time about a dozen years ago when Harry Stubbs (who writes first-class science fiction under the pseudonym of Hal Clement) learned that I had never looked through a telescope. Saddened by this, he took me to Milton Academy, where he taught science, and had me look through a telescope which he had focused on the Moon for my benefit.

Abandoning my amateur status, I looked through a telescope for the first time in my life, stared at the Moon, then stared at Harry with a wild surmise, and said, "Good heavens, there *are* craters on the Moon."

Years later I repeated this triumph of observation—

On June 30, 1973, I was on the good ship *Canberra*, a hundred miles or so off the coast of West Africa, observing a total eclipse of the Sun—the first total eclipse I had ever seen.

My excitement was extreme, and since I was standing in a forest of cameras, telescopes, and tape recorders, set up by hundreds of amateur astronomers, someone was bound to record my comments, and someone did. That someone then played back the tape to me with great amusement.

Aside from incoherent shouting, two clear statements could be heard. First, I expressed approval of the spectacle as compared with the photographs I had seen. "Yes," I cried, over and over, "that's the way it's supposed to be."

Then I made a clever deduction. "The stars are out," I cried, "and that *proves* the stars are there in the daytime."

Well, in addition to that important scientific finding, I enjoyed the eclipse very much; it lived up to all my expectations (except that it was twilight that fell and not night) and, as a result, I am going to talk about eclipses here.

Whenever a dark body is in the vicinity of a source of light, that body casts a shadow. We know that as a matter of common knowledge, but most of us don't stop to think that the shadow is a three-dimensional phenomenon. What we commonly refer to as the shadow is only a cross section, a two-dimensional darkening of a surface. If, however, the surface is moved toward or away from the light source, it remains in the shadow, which is thus shown to be three-dimensional.

The shape of the shadow depends on the shape of the object casting the shadow, and with respect to astronomical bodies the case is simple. The opaque bodies that most concern us are spheres and therefore the shadows are conical in shape.

If the opaque body is relatively small compared to the source of illumination and relatively close to it, the cone converges to a point at not too great a distance from the body. This is the case with the Earth and the Moon, and with the shadows they cast on the side opposite that of the illuminating Sun.

In the case of the Earth, the shadow (which we ourselves enter every evening after sunset and which we call "night") narrows to a point at a distance of 860,000 miles for Earth's center in the direction, of course, exactly opposite that of the Sun. At a distance of 238,800 miles from the Earth's center (the average distance of the Moon from the Earth) the narrowing shadow has a circular cross section with a diameter of 5,800 miles, as compared to Earth's own diameter of 8,000 miles.

The Moon has a diameter of 2,160 miles, so that when it passes through Earth's shadow, as it sometimes does, it passes through a shadow that is 2.7 times as wide as itself. The entire Moon can be darkened by that shadow, therefore, in a "lunar eclipse" and can remain in total shadow for as long as an hour and a half.

The Moon, which is as far from the Sun, on the average, as the Earth is, casts a shadow which narrows at the same rate that Earth's does. Since the Moon starts off with a considerably smaller diameter than Earth does, the narrowing brings the lunar shadow to a point at a correspondingly closer distance to its source. The Moon's shadow comes to a point at a distance of 234,000 miles from the Moon's center.

The Earth's center is, on the average, 238,800 miles from the Moon's center. That portion of the Earth's surface which is directly under the Moon, being 4,000 miles above the Earth's center, is 234,800 miles from the Moon's center.

It follows, then, that if the Moon is directly between the Earth and the Sun, the Moon's shadow comes to a point and ceases nearly one thousand miles *above* Earth's surface. It does *not* reach the Earth.

Still, it seems odd that the point is so close to Earth's surface. Is this really unusual in the sense that it is pure coincidence, or is there some compelling astronomical reason? Let's see.

If the point of the Moon's shadow reached the surface exactly, that would be equivalent to saying that the apparent size of the Moon and the Sun in the sky would be precisely the same. And indeed to the naked eye this seems to be so. The Sun and the Moon have always been taken to be equal in size, if not in brightness, and probably this has seemed natural. If you have two lamps in the sky (in early days it was taken for granted that the only purpose of the Sun and the Moon were to supply light for the all-important Earth), why shouldn't they be the same size, regardless of brightness? To put it in modern terms, are you surprised that a 100-watt bulb and a 60-watt bulb both have the same physical dimensions?

Yet if we measure the apparent diameter of the Sun and the Moon carefully, we find that the sizes are not, indeed, precisely equal. The average apparent diameter of the Sun is 1,919 seconds of arc, while that for the Moon is 1,865 seconds of arc.* The fact that the Moon, in appearance, is actually a bit smaller than the Sun is equivalent to the statement that the point of the Moon's shadow falls a bit short of the Earth's surface.

So now we need ask why the Moon and the Sun are nearly equal in apparent size.

In actual fact, the Sun is far larger than the Moon. The Sun's

* Sixty seconds of arc equals 1 minute of arc and 60 minutes of arc equals 1 degree.

diameter of 864,000 miles is nearly precisely 400 times larger than the Moon's diameter of 2,160 miles. Of course, the apparent size depends not only on the actual size of the two bodies but also on their respective distances from the Earth. If the Sun were exactly 400 times as distant from us as the Moon, the disparity in diameter would be exactly balanced and both bodies would appear to be the same size.

The average distance of the Moon from the nearest portion of the Earth's surface is 234,800 miles while the average distance of the Sun is 92,900,000. The Sun, in other words, is 395 times as far from us as the Moon is and the difference in real size is nearly balanced in consequence. Because the Sun is not *quite* 400 times as far away as the Moon is, it is a little larger in apparent size than the Moon is and the point of the Moon's shadow ends just a little short of the Earth's surface.

But you know there is nothing that compels either the Sun or the Moon to be exactly the sizes they are or either to be at exactly the distances they are. The fact that the ratio of sizes nearly matches the ratio of distances is pure coincidence—and, of course, a very lucky coincidence for us.

It is not even an enduring coincidence. The Moon has not always been at its present distance from the Earth, nor will it always remain there. Because of tidal action, it is very slowly receding from the Earth. It has been much closer to Earth in the past (at which time its swollen body appeared markedly larger than the Sun), while it will be much farther from Earth in the future (and then its shrunken body will appear markedly smaller than the Sun.)

But let's get back to the Moon's shadow. The narrowing cone of shadow that I have been talking about is called the "umbra" from a Latin word for "shadow." Surrounding the umbra on all sides, we can picture another diverging cone that starts at the Moon and gets wider and wider as it moves away from the Moon. This is the "penumbra" ("almost shadow").

It is called that because if you imagine yourself located at any point within the penumbra, you will see the Moon cover part of the Sun. That is, you will see a partial eclipse. (The word "eclipse" is from a Greek word meaning "to omit." During a solar eclipse, part or all of the Sun is "omitted" from the sky.)

The closer your location within the penumbra to the umbra itself, the greater the fraction of the Sun you will see covered by the Moon. Within the umbra, *all* the Sun is covered; and to a very real extent that is all that really counts. Let me explain why—

If you are standing on the Earth's surface and the Moon's penumbra, but *not* the umbra, passes over you, then what you would see if you look at the Sun (*which you should not*) would be the Moon skimming by the Sun, obscuring more and more of it to a maximum that is less than total and then obscuring less and less of it till the Moon moves away altogether. The maximum obscuration can be any amount from barely more than 0 per cent to barely less than 100 per cent, depending on how close the umbra comes to you.

In most cases of this sort, it seems very unlikely to me that you would know anything had happened if you hadn't been warned by the newspapers (which were in turn warned by the astronomers) that something was going to happen.

For one thing, no one looks at the Sun (or *should look at it*) for more than a second or so voluntarily, and when one does all one sees is a formless and blinding blaze. Even if the Moon covered up half the Sun and left only a fat crescent of luminosity, you would not see that in the quick second of glance—you would still see only the usual formless dazzle.

I know that in my own case on June 30, 1973, it was impossible for me to see, with the bare eye, the details of what was happening to the Sun as long as the tiniest bit of luminous surface remained exposed.

To be sure, I could see the advance of the Moon every time I looked at the Sun (briefly!) through a piece of very darkened plastic supplied us by the cruise organizers,† but I would never have done this if I had not known in advance there would be something to see.

I am convinced, then, that primitive peoples were not usually frightened by ordinary partial eclipses, simply because they would not be aware anything was taking place.

Ordinary partial eclipses, that is. There can be exceptional conditions. There might be, for instance, a haze which happens to dim the Sun to the point where it can be looked at without harm and

† These were chiefly Marci Sigler, her husband, Phil Sigler, and her brother Ted Pedas, who did a marvelous job and to whom I am very grateful.

yet not so thick as to obscure it totally—a kind of natural sheet of plastic in the sky. In that case, the encroaching circle of the Moon would be all too plain and seeing it would be unavoidable. Such a combination of rarities, however, as a partial eclipse and just the right haze can only happen once in many centuries.

Even in a clear sky, a partial eclipse, if it is close enough to total-ity, can produce noticeable effects on the objects about us. If the partial eclipse succeeds in obscuring more than 80 per cent of the surface of the Sun, the quality of sunlight changes. It begins to fade and take on a claylike, dead look, which is like nothing else on Earth. After all, there is nothing in the sky of the brightness of a fraction of the Sun. The unobscured Sun itself is much brighter and everything else is much dimmer.

The light on June 30, 1973, made me uneasy though I knew what was happening, and from the remarks I heard all about me, even the experienced eclipse buffs were affected. I can well imagine that in such a light, primitive people, who might be unaware that an eclipse was taking place, would nevertheless find themselves feeling that the Sun must be flickering out and would begin trying to look up at it in fear and apprehension.

What's more, if you happened to be in a wooded area where the sunlight dappled its way between the leaves, you would be accus-tomed to seeing the sunlight on the ground as a series of overlap-ping circles. This is because the tiny spaces between the leaves serve as pinhole cameras so that each circle of light is actually an image of the Sun.

Well, as the Moon cuts off part of the Sun, each one of the over-lapping circles of light becomes a little chopped off where the Moon is encroaching. What with the shifts imposed by the move-ment of the leaves, this would not be casually noticeable at first. By the time, however, that the Moon's shadow had cut off 80 per cent or more of the Sun, what was left would be a thinnish crescent, and under the tree there would be a myriad dancing crescents in pale and sickly light—and that too would lend an aura of abnor-mality to the situation.

So primitive people might grow increasingly uneasy, even though they did not know what was happening and could not tell by look-ing at the Sun what was happening. Yet if the light dimmed but did not go out altogether, it seems to me that outright ponic would be averted, for the light would start to brighten again soon enough. The partial eclipse might last as much as two hours altogether, but it would only be during the fifteen minutes at the center of the period that the effects would be noticeable.

It is when the eclipse is actually total; when the light dies down to a mere twilight and the stars come out; when one can finally look at the Sun itself and see only a dark circle (surrounded by a comparatively dim corona, to be sure) where it used to be; *that* is when panic would strike.

After all, a primitive observer would not know that the dark circle was the Moon obscuring the Sun. It would be only natural to think that the dark circle was the Sun itself, with its fires out—and there's not one of us that wouldn't be afraid to the point of panic if we thought that were so, and that we had to face a world of cold and darkness forever. (I can imagine a primitive observer's relief when the first bit of Sun emerged from behind the Moon so that the dark circle seemed to be reigniting and catching fire again.)

I know that there are all kinds of myths about dragons or other monsters swallowing the Sun at the time of eclipse, and we hear tales of primitive people coming out to make noise that will frighten the monster away and force it to let go of the Sun—but I can't help thinking that this is already a case of advanced thought.

To suppose that the Sun is overtaken by something, even if that something is a dragon rather than the Moon, and to suppose that the Sun is intrinsically unharmed and will shine as always once the dragon lets go—is already good astronomy.

No, I think it much more likely that men, at the very beginning, could only think that the Sun's fires were dying, as a campfire might; and that, at the conclusion of the eclipse, the Sun reignited as a smoldering campfire would on the introduction of new fuel.

Since it is only the total eclipse that is of consequence, then, to all but the highly sophisticated, and since the phenomena accompanying the total eclipse are widely different and far more spectacular than anything accompanying even the most nearly total of partials, let's stick to the total from now on. That means we must imagine ourselves inside the Moon's umbra.

From any point within the umbra, we see the Moon cover the entire face of the Sun. If we imagine ourselves somewhere on the central axis of the umbra—on the mid-line of the shadow, extending from its point to the central point of the night side of the Moon—then that central point of the Moon's visible circle is exactly over the central point of the Sun's visible circle. The Moon laps over the Sun, so to speak, by an equal amount in every direction.

If we move along this mid-line closer and closer to the Moon, the apparent size of the Moon increases and it laps over the Sun to a greater extent in every direction. When we actually reach the Moon, that body obscures half the sky and the other half is all shadow—it is nighttime on the Moon.

(To be sure, as we move closer and closer to the Moon, we also move closer and closer to the Sun, so that the Sun's apparent size also increases. However, the Sun is so much farther away from us than the Moon is that our approach is a far smaller percentage of the Sun's total distance, and the Sun's increase in apparent size is very small—so small that it can be ignored.)

If we move along the mid-line farther and farther away from the Sun, the apparent size of the Moon decreases (again to a far greater extent than the apparent size of the Sun decreases) and it laps over the Sun to a lesser extent in every direction.

If we move along the mid-line of the shadow to a place sufficiently far from the Moon, we eventually reach the apex of the cone, the point to which it dwindles. Seen from that point, the Moon is exactly the same size as the Sun, the central points of the two bodies coincide, and the Moon fits *exactly* over the Sun.

But at this apex, we are still well above the surface of the Earth if we go by what I said earlier in the article. What if we extend the mid-line of the cone past the apex, however, and imagine ourselves dropping down that mid-line to the surface of the Earth. What do we see then?

Once past the apex of the cone of the umbra, the apparent size of the Moon is less than that of the Sun. The Moon does not cover the Sun and it is the Sun that laps over. Assuming the central points of both circles in the sky coincide, as they would if we remained on the extension of the mid-line, the Sun laps over an equal amount in every direction.

What we would see in that case is a different kind of partial eclipse, one in which a thin section of the outermost portion of the solar disc gleams like a luminous ring about the black center that is the obscuring Moon. This is called an "annular eclipse" from the Latin word for "ring."

An annular eclipse of the Sun, being partial, has none of the phenomena associated with it that a total eclipse has. If, as I have said earlier in this chapter, the point of the Moon's shadow is always a thousand miles or so above the surface of the Earth, an annular eclipse is all we can ever see when the Moon passes squarely in front of the Sun.

How, then, is it that I have talked about total eclipses? How is it that I myself saw a total eclipse on June 30, 1973?

If, indeed, the Earth moved around the Sun in a perfect circle, remaining always 92,900,000 miles from its center, and if, indeed, the Moon moved around the Earth in a perfect circle, remaining always 238,800 miles from center to center, then an annular eclipse would indeed be all we would ever see, for the apex of the umbra would always skim the Earth a thousand miles above its surface.

But the Earth moves around the Sun in an elliptical orbit, with the Sun at one focus of that ellipse and therefore off-center. This means that the Earth is sometimes closer to the Sun and sometimes farther. The Sun's average distance from Earth is, indeed, 92,900,000, but at the point in its orbit when Earth is farthest from the Sun ("aphelion") the distance is 94,500,000 miles. At the opposite point in its orbit, the Earth is closest to the Sun ("perihelion") and the Sun is only 91,500,000 miles away.

This reflects itself in the apparent size of the Sun; not much, to be sure, since the total difference in distance is only 3.2 per cent, so that the Sun looks about the same size all through the year. Careful measurement, however, shows that although the Sun's average apparent diameter is 1,919 seconds of arc, it is fully 1,950 seconds of arc in diameter at perihelion and only 1,888 seconds of arc at aphelion.

The situation for the Moon is similar. The Moon moves in an elliptical orbit about the Earth, with the Earth at one focus of the ellipse and therefore off-center. Although the average distance of the Moon's center from the Earth's center is 238,800 miles, it passes through a maximum distance ("apogee") of 252,700 miles and a minimum distance ("perigee") of 221,500.

This is a 13 per cent difference and reflects itself in a change in the Moon's apparent diameter that is greater than that of the Sun (though still not great enough to impress the casual observer).

Although the apparent diameter of the Moon is, on the average, 1,865 seconds of arc, it reaches a maximum diameter of 2,010 seconds of arc and a minimum of 1,761 seconds of arc.

As the distance of the Sun from the Earth (and therefore from the Moon) changes, the slope by which the cone of the Moon's shadow narrows also changes. The farther the Sun is from the Earth-Moon system, the more gradually the umbra narrows and the farther its point from the Moon—and therefore the closer the point can come to the Earth's surface.

And, of course, when the Moon is closer than average to the Earth, the point of the umbra moves with it and is also closer to the Earth. If the Sun is far enough from the Earth and the Moon is close enough, then the point of the umbra actually reaches the surface of the Earth. Indeed it could penetrate beneath the surface if we imagined the structure of the Earth to be transparent. That is why total eclipses are possible.

If an eclipse should take place just at the time when the Sun was at its farthest and the Moon at its closest to Earth, then the width of the umbra at Earth's surface would be 167 miles. In that case, a little over 21,000 square miles of the Earth's surface (half the area of New York State) could be in the Moon's shadow at a particular instant of time.

Still, the *average* position of the point of the umbra is well above the Earth's surface, so that it follows that more than half the solar eclipses seen from Earth's surface would, when seen under the most favorable conditions, be annular rather than total. Too bad!

Let's look at this in another way.

Instead of worrying about the distances of the Moon and Sun, let's just consider their apparent size. The Moon can be anything from 1,761 to 2,010 seconds of arc, and the Sun anything from 1,888 to 1,950 seconds of arc.

The smaller the Moon in appearance and the larger the Sun, the more the Moon misses out in any effort to cover the whole of the Sun. When the Moon is at its smallest and the Sun at its largest and the central points of both circles in the sky coincide, a 94-second-of-arc width of gleaming Sun shines out beyond the Moon in every direction. The Moon succeeds in covering only 82 per cent of the Sun.

On the other hand, when the Moon is at its largest in appearance, and the Sun is at its smallest, not only does the Moon cover the entire Sun but, when they are center to center, the Moon laps over the Sun in every direction by 61 seconds of arc.

This overlap isn't terribly important in one way. If the Moon fits exactly over the Sun, then all the phenomena of a total eclipse are visible. The fact that the Moon further overlaps the Sun by a certain amount in no way improves the phenomena. Rather it tends to diminish them by a bit since events immediately adjacent to the Sun's surface are obscured.

Nevertheless, the overlap becomes important if we take time into consideration. After all, the Sun and Moon are not motionless in the sky. Each is moving west to east relative to the stars, the Moon at a rate thirteen times that of the Sun so that it overtakes and passes the Sun.

Hastening after the Sun, the Moon's eastern edge makes contact with the Sun's western edge for "first contact," and the partial phase of the eclipse begins.

The eastern edge of the Moon encroaches across the face of the Sun and, assuming the Moon is passing squarely in front of the Sun, in just about an hour that eastern edge of the Moon reaches the easternmost edge of the Sun to make "second contact." The last light of the Sun shines like a glorious diamond at one end of the dark circle of the Moon. The "diamond-ring effect" fades quickly and the corona comes out. (When I watched the eclipse I held the polarized film before my eyes just a bit too long and I did not see the dying diamond.)

When the western edge of the Moon reaches the western edge of the Sun, we have "third contact." At this point the Moon has passed by the Sun and the western edge of the Sun peeps out from behind the Moon forming a diamond-ring effect again. This time the diamond waxes rapidly in brilliance and in a few seconds it is too bright to look at. (I watched this, and those two or three seconds were, for me, the most spectacular and wonderful of the eclipse. An involuntary cry of appreciation came from everyone watching. —How much more heart-felt if we were all primitives who thought the Sun had actually gone out!)

With third contact, the total eclipse is over and the partial eclipse is on again, as the Moon slowly leaves the Sun. Finally the western edge of the Moon is at the eastern edge of the Sun and with this "fourth contact," the eclipse is over.

In the case of an annular eclipse, the third contact occurs *before* the second contact. In other words, the western edge of the Sun is revealed before the eastern edge is obscured and there is never true totality.

In a total eclipse, where the point of the umbra just reaches the surface of the Earth, the Moon fits exactly over the disc of the Sun and the second and third contacts occur simultaneously. The Sun is no sooner totally covered when the western edge is uncovered and the total eclipse is over.

The further within the umbra the Earth's surface is, the larger the Moon is in comparison with the Sun, and the farther the Moon must travel between second contact and third. When the Moon is at its largest and the Sun is at its smallest, and when second contact is then made, the Moon has just barely covered the eastern edge of the Sun while its western edge still laps beyond the western edge of the Sun.

The Moon must then travel for some time before its western edge reaches the western edge of the Sun, makes third contact, and brings totality to an end. Seen from any one point on the Earth's surface, the maximum length of time that totality can last under the most favorable conditions is a little over seven minutes.

The eclipse that I viewed from the *Canberra* was quite long and at its best could be viewed for nearly seven minutes. From the actual spot at sea where the ship was located, totality lasted for nearly six minutes. There won't be another as long for over a century.

So all things considered, I'm glad I saw it, even though it did mean I had to travel a total of just over 7,000 miles (I hate traveling) and even though it did mean I had to stay away from my typewriter for fifteen days.

I'm so glad I saw it that I plan to continue talking about eclipses in the next chapter.

6.   Dance of the Luminaries

Very much of my life is spent in trying to decide how best to explain rather complicated phenomena in such a way that people can understand without suffering too much pain in the process.

There are times when I feel sorry for myself for having to work at this task, and I worry about the strain imposed on the delicate structure of my brain. Whenever that happens, though, I remember that I am dealing with an audience that *wants* to understand and is anxious to meet me halfway. There then follows an upwelling of gratitude for all my Gentle Readers that drowns any feeling of self-pity I may have been experiencing.

After all, what if my life-style had forced me into missionary endeavors designed to convert *hostile* audiences. Surely I would get nothing but beatings for my pains.

For instance, I will not conceal from you my lack of sympathy for the tenets of Christian Science. I'm sure Christian Scientists are fine people and I would not for worlds interfere with their happiness, but I cannot make myself accept their denial of the reality of the phenomenal or material world. I don't believe that disease, to take an example, is not real and that it can be removed by faith or prayer.*

Consider, then, that immediately across the street from the apartment house in which I live is a Christian Science church. And consider further that on Sunday mornings in the summer there is a steady menacing roar that I can hear even through closed windows. When I first became aware of this, I asked questions in alarm and was told that it was the fan of the large air conditioner used by the said church.

* Yes, I know cures have been reported. Cures have been reported for every conceivable kind of treatment. But what's the over-all batting average?

Naturally, I laughed loud and long at the thought that the congregation should incorporate into the architectural structure of their house of worship a device that so thoroughly refuted everything they had to say. If the material world is not real and if the faithful Christian Scientist can by prayer eliminate disease, cannot a whole churchful of them pray away the sensation of heat? Do they need to bow to the material by installing an air conditioner?

But I laughed also out of relief that I felt no impulse to go into the church and lecture them on this. Let it go! I would but get myself smashed into kindling if I tried.

I am so glad that my impulse, instead, is to follow up the previous chapter by explaining to you the reason why there are so few eclipses of the Sun, and how it was that the ancients could predict them.

For there to be a solar eclipse, the Moon has to be in front of the Sun, as viewed from the Earth, so let's begin by considering the relative positions of the Moon and the Sun as we see them.

As the Moon revolves about the Earth, it is sometimes on the same side of us that the Sun is, and sometimes on the opposite side. When the Moon is on the opposite side of the Earth from the Sun, sunlight goes past the Earth to shine on the Moon. If we look at the Moon, then, with the sunlight coming over our shoulder toward it, so to speak, we see the entire side of the Moon, facing us, to be illuminated. We see a complete circle of light and are looking at the "full Moon."

If, on the other hand, the Moon is on the same side of the Earth that the Sun is, it is between the Sun and the Earth (since it is much closer to us than the Sun is) and we don't see it at all, for it is in the full glare of the Sun. Even if we could somehow block out that glare, we wouldn't see the Moon anyway because it would be the side away from us that would be receiving the light of the Sun. The side toward us would be dark.

When the Moon moves a little to the east of the Sun (as it progresses west to east in its orbit about the Earth) we begin to see a bit of the sunlit side. This we see as a narrow crescent, but only when the Sun is not in the sky. After the Sun has set, the crescent Moon is briefly visible in the western sky at twilight before it, too, sets. Ancient man thought of it as a "new Moon" ready to go through another month-long cycle of phases, and that phrase has come to be applied to the Moon when it is between us and the Sun.

As the Moon moves around the Earth, the visible crescent grows steadily until we see a full Moon and then the lighted portion shrinks steadily until it vanishes and we have a new Moon again.

The time from one new Moon to the next is a little less than a calendar month (I'll get into exact figures later), so that in the space of a little less than a year there are twelve new Moons. About one year in three, the first new Moon comes so early in the year that there is time for a thirteenth new Moon before the end of the year.

If we have a solar eclipse only when the Moon is in front of the Sun, the eclipse must come at the time of the new Moon, for it is then that the condition is fulfilled. In that case, though, why isn't there an eclipse of the Sun at *every* new Moon? Why aren't there twelve solar eclipses every year, and one year out of three thirteen of them?

To see the answer to that, let's approach the problem from a different direction.

The Earth goes around the Sun in 365.24220 days. As it moves in its orbit, we see the Sun against a progressively different part of the starry background.† Therefore, it seems to us that the Sun, moving steadily west to east, makes a complete circle of the sky in 365.24220 days. This circular path, which the Sun seems to mark out against the sky in the course of the year, divides the sky into two equal halves and is therefore called a "great circle." The particular great circle the Sun follows is the "ecliptic."

(The Sun, along with everything else in the sky, appears to move east to west, making a circle in twenty-four hours, thanks to Earth's west-to-east rotation on its axis. This daily rotation does not concern us at the moment. If we consider the motion of the Sun *relative to the stars*, the effect of the Earth's rotation is eliminated.)

The Moon moves about the Earth in 27.32 days, relative to the stars. For this reason, the 27.32-day period is called the "sidereal

† Actually, when we look at the Sun we can't see any stars in the sky; they're drowned out by scattered sunlight. Little by little, though, we can plot the entire starry sphere of the sky since in the course of a year all of it becomes visible at one time or another. If we note an appropriate point at midnight on any night of the year, we know that the Sun is at the directly opposite point in the sky and know its position against the stars there.

month," from a Latin word for "star." As a result, we see the Moon also make a west-to-east circle in the sky relative to the stars. The path followed by the Moon in the sky is also a great circle and also divides the sky into two equal halves.

A complete circle is divided into 360 equal parts, called "degrees" and symbolized °. If the Moon moves about its circle west to east at a constant speed (it doesn't, but it *almost* does, so let's accept the constancy for the sake of simplification), then each day it moves 360/27.32, or about 13.2°, eastward relative to the stars. The Sun, moving through its circle in 365.24220 days moves eastward at a rate of 360/365.24220, or 0.985°, per day.

Now imagine that the Moon and Sun are in the same spot in the sky, so that the Moon is directly between us and the Sun, and it is the time of the new Moon. If the Moon moves eastward, making a complete circle in the sky, and returns to that same spot (relative to the stars) in 27.32 days, is it new Moon again?

No, for in those 27.32 days, the Sun has slipped 27° eastward and is no longer in that same spot. The Moon must continue in its eastward journey in order to catch up with the Sun and it takes nearly two and a quarter days for it to do so. For this reason, the time from one new Moon to the next is, on the average, 29.530588 days.

It is the new Moon that has religious significance and is of concern to ecclesiastical assemblies, or "synods." The period from one new Moon to the next is, therefore, the "synodic month."

If the Moon's circle in the sky coincided with the ecliptic—that is, if both Sun and Moon traveled exactly in the same path in the sky— there would indeed be an eclipse of the Sun at every new Moon.

There is no compelling reason, however, why the Moon should follow the precise path of the Sun, and it doesn't. The two paths, that of the Sun and that of the Moon, are great circles that do not coincide.

Two great circles in the sky that do not coincide are compelled, mathematically, to cross each other at two points, one of which is at a place in the sky directly opposite the other. These crossing points are called "nodes," which is from a Latin word for "knot," since a knot between two lines always comes at the point where they cross.

At the point where the great circle of the Moon's path crosses that of the Sun's path, an angle of 5.13° is marked out. You can say this in more formal language thus: The Moon's orbital plane is inclined 5.13° to the ecliptic.

This really isn't much. Suppose you imagined a balloon a hundred feet across and draw two great circles around it with the plane of one inclined 5.13° to the other. They would meet at two points on opposite sides of the balloon, of course, and between these nodes, the two circles would separate and come together again. The two places of maximum separation would come exactly halfway between the nodes and the separation would there be 5.13°.

On a balloon a hundred feet in diameter, such a separation would be 4.5 feet out of a total circumference of 314 feet. If a bacterium in the center of the balloon could see and consider those two lines and find them to be separated by never more than 4.5 feet, it would surely feel that even though they did not coincide, they were both in the same general part of the balloon.

In the same way, though the paths of the Sun and the Moon do not coincide exactly, they are clearly in the same general part of the sky. The path followed by the Sun is divided among the twelve constellations of the zodiac, and the path followed by the Moon is also to be found in those constellations.

Yet from the standpoint of eclipses, this inclination of the Moon's path to the ecliptic is important.

At every new Moon, the Moon passes the Sun west to east, and it is then as close to the Sun as it can get in that particular go-round of the dance of the luminaries. If the Moon passes the Sun at the point where their two paths are at their greatest separation (midway between the two nodes) there is then a gap of 5.13° between the center of the Moon and the center of the Sun.

Since the apparent radius of the Sun (the distance from its center to its edge) is, on the average, 0.267° and that of the Moon is 0.259°, the distance between the edge of the Sun and the edge of the Moon is 5.13—0.267—0.259, or 4.60°. Another way of saying it is that the gap between the edges of the Moon and the Sun is great enough at this time for something like nine bodies and the size of the full Moon to be placed across it side by side.

Naturally there would be no eclipse under such circumstances since the Moon is *not* exactly between us and the Sun. Indeed, if we could see the Moon at this time and somehow ignore the glare of the Sun, we would see a thin crescent on the Moon's sunward side. We can't ever see it actually, though, since at this time, the Moon rises and sets at virtually the same time as the Sun and we never get a chance to look at it in the Sun's absence.

Of course, if the Moon happened to be passing the Sun at a point where the two were nearer the node, the gap between them would be smaller. If the Moon passed the Sun at a point where both were exactly at either of the nodes, then, indeed, the Moon would move exactly in front of the Sun and there would be an eclipse. (This eclipse would be either total or annular according to the principles discussed in the previous chapter.)

The eclipse would be visible over only a portion of the Earth. The path of totality (or annularity) would be a line crossing the center of that side of the Earth's globe facing the Sun-Moon combination and would therefore be in the tropic zone. The exact latitude would depend on the time of the year; in our winter the southern half of the tropics would get it, and in our summer the northern half would. The exact longitude would depend on the position of the Earth with respect to its rotation about its axis.

All around the line of totality, there would be a region where people could see a portion of the Sun peeping out from behind the Moon throughout the course of the eclipse. For them only a partial eclipse would be visible. Still further from the line, the vantage point would be such that the Moon would seem to miss the Sun entirely and for people in those regions of the Earth there would be no visible eclipse at all.

Let's count everything as an eclipse, however, if it is visible anywhere on Earth even partially.

The next question, then, is whether the Moon has to pass the Sun *exactly* at the node to produce an eclipse.

No, it doesn't. Suppose the Moon passes in front of the Sun a small distance from one of the nodes. In that case, the Moon passes in front of the Sun but not dead center. As viewed that that portion of the Earth's surface directly under the Sun-Moon combination, the Moon will seem to pass the Sun a little north of center or a little south of center. The shadow will then cross the Earth a little north or a little south of its mid-line (as seen from the Moon). The temperate zones will then get a chance to see an eclipse, or even the polar zones.

The farther from a node it is that the Moon passes the Sun, the farther north, or south, the total eclipse can be seen. Eventually, if the passage takes place far enough from a node, the line of totality misses the Earth altogether, but a partial eclipse can still be seen either well to the north or well to the south. And if the passage takes place still farther from the node, even the partial eclipse can't be seen anywhere.

On the average, a total or annular eclipse of the Sun will be seen somewhere on Earth if the Moon passes the Sun within 10.5° of a node, and at least a partial eclipse will be seen somewhere on Earth if the passage takes place within 18.5° of a node.

Next, let's think about those nodes for a while. During half its circuit, the Moon is north of the ecliptic and during the other half south of the ecliptic. At one node, the Moon moves across the ecliptic at a shallow angle from north to south. This is called the "descending node," from the long-established Western habit of putting north at the top of a map and south at the bottom. At the other side of the orbit, the Moon moves across the ecliptic from south to north and that is the "ascending node."

As it happens, these nodes do not stay in the same place relative to the stars. For a variety of gravitational reasons that we don't have to go into, they move (in step, of course) around the ecliptic, east to west, making one complete turn in 18.6 years.

If the nodes were to stand still relative to the stars, the Sun would make a complete circuit from ascending node back to ascending node, or from descending node back to descending node, in exactly one year. As it happens, though, the nodes move toward the approaching Sun so that the meeting takes place sooner than would be expected. The approach of the nodes cuts the time by nearly nineteen days, so that the time it takes the Sun to make a complete circuit from ascending node back to ascending node, or from descending node back to descending node, is 346.62 days. Let's call this the "eclipse year."

The Sun's motion across the sky, *relative to either of these nodes* is a little faster than its motion relative to the motionless stars.

Whereas the Sun moves at 360/365.24, or 0.985°, per day relative to the stars, it moves 360/346.62, or 1.07°, per day relative to the nodes. In either case, it is moving eastward, of course. In the course of a synodic month, the Sun moves eastward 1.07×29.53, or 31.60°, relative to the nodes.

This means that every time the Moon reaches the new Moon position it is 31.60° closer to one of the nodes than it was the time before. Eventually it will approach the neighborhood of the node and the question arises as to whether it will pass the Sun too far short of the node one time and then too far past the node the next time so that on neither occasion will there be an eclipse.

For that to happen, the Sun would have to be farther than 18.5° west of the node the first time and farther than 18.5° east of the node the next time. It would have to move at least 37° during the time between new Moons—*and the Sun can't do that*. Relative to the nodes, the Sun can only move 31.6° in the gap between new Moons. If one new Moon takes place just short of the 37° stretch, the next one will inevitably find the Sun within it.

This means that there has to be an eclipse—total, annular, or partial—visible from some place on Earth, *every single time the Sun is in the neighborhood of either node*. This, in turn, means that there must be *at least* two solar eclipses every year.

Then, of course, if the Sun happens to be slightly within the eclipse region surrounding the nodes at one new Moon, it will not quite have cleared the region by the time of the next new Moon, so that there will be two solar eclipses on Earth in successive months. That means three eclipses for that year or, if it happens at both nodes, as it may, four eclipses.

Remember that the Sun makes a complete circuit from one node back to that same node in 346.62 days, which is nineteen days less than a calendar year. That means if it passes through one node very early in the year, it will have time to reach that same node before the year is out. It can then pass through two eclipses at one node, two at the other, and return for another one at the first node. In this case, there can be *five* eclipses in a single year, as in 1971, and this is a maximum.

Under the most favorable circumstances, six successive eclipses must stretch over a period of 376 days, so there cannot be six eclipses in a year.

There it is, then. In any given year, there are anywhere from two to five eclipses of the Sun and this sounds like quite a lot, but don't be fooled. They are invariably in different places on Earth, so you can't expect to see all of them without traveling. The chances, alas, are that in any given year you won't see any of them without traveling.

In fact, if we restrict ourselves to total eclipses, the average length of time between such phenomena for any given spot on Earth is 360 years. This means that if we take the average human life expectancy to be seventy-two years, we can say that a particular person has only one chance in five of seeing a total eclipse in his lifetime if he doesn't travel.

Now what about predicting eclipses?

Imagine the Moon passing the Sun just at a node. This starts a particular pattern of eclipses that can be (and is) quite complicated, because the next time the Sun is at a node, the Moon isn't. Eventually, though, the Moon once again passes the Sun just at that node and the whole pattern starts all over again. If we learn the pattern for one whole cycle, we can predict the eclipses in the next cycle even without a lot of fancy astronomic sophistication.

But how long will it be between successive appearances of Sun and Moon precisely at the same node? Clearly this will come at a time when there has been an exact number of synodic months in the interval and *also* an exact number of eclipse years in the interval. In that case, both Sun and Moon are back where they started.

Unfortunately, the two periods, that of the synodic month and that of the eclipse year, are not easily commensurate. That is, there is no small figure that will be at once a whole number of each period within the limits of astronomic accuracy.

There are numbers, however, that are fairly small and that are *almost* correct.

Consider nineteen eclipse years, for instance. That comes out to 19×346.62, or $6,585.78$, days. Next consider 223 synodic months. That comes out to 223×29.53058, or $6,585.32$, days.

Every $6,585+$ days, then, both the Sun and the Moon have come back to the same place with respect to the nodes, so that if there was an eclipse the first time, there will be an eclipse the second time also, and then a third eclipse 6,585 days later still and then a fourth, and so on.

This nineteen-eclipse-year period is called the "saros," a numerical term the Greeks borrowed (with distortion) from the Babylonians, who were probably the first to discover it.

The correspondence isn't exact, of course, since there is a 0.46 day difference. This means that at the end of the nineteen-eclipse-year period, the Moon has not quite caught up with the Sun. This means that each succeeding eclipse of the series takes place a little farther back with respect to a node. If it is the descending node, each succeeding eclipse of the series falls a little farther north on the Earth's surface; if it is the ascending node, a little farther south.

The whole series of eclipses that takes place for a given position of the Moon and Sun with respect to the descending node, at 6,585-day intervals, begins far in the south as a partial eclipse, progresses farther and farther north, becoming more nearly total until it finally becomes completely total (or annular) in the tropics; then continues still farther north, gradually becoming only partial once more and fading out in the extreme north. The same happens for eclipses at the ascending node except that the progression is then southward. Altogether, eighty-one eclipses may be seen in a particular series over a period of about 1,460 years.

Because 223 synodic months comes to 6,585.32 days, there is a complication. There is that extra third of a day which gives the Earth a chance to rotate before the eclipse takes place. Each successive eclipse in the saros series marks its path about eight thousand miles farther west than the one before.

This means that every third eclipse in the series is back where it was, just about, on the surface of the Earth, except that it is farther north, or farther south, depending on the node.

One third of the eclipses of the series, twenty-seven, fall in about the same longitude and several of them will be seen from some given point as partials of varying degree. At most, only one of them will be a total eclipse as seen from some given point.

Once the saros was discovered, it was easy to take note of all the eclipses that took place at a given region of the world and realize that each forms part of a series and will more or less repeat itself every fifty-seven eclipse years (or every fifty-four calendar years) for a while.

If you note that a particular series has been fading out and at its last appearance was only slightly partial, you'll know it won't appear again. If, on the other hand, the eclipse has been covering more of the Sun at each appearance, there is a chance that at the next appearance it may be nearly total, or even quite total.

The first person we know of by name who, in the Western tradition, predicted that an eclipse would take place in a particular year was Thales of Miletus, who was born about 640 B.C. on the Asia Minor coast. He had visited Babylonia and had probably studied the eclipse records that the astronomers there had been keeping for centuries. Noting that there had been a pretty fat partial eclipse in the year we now call 639 B.C., which had been visible in western Asia, he predicted there would be an even better one in 585 B.C.

There was. A total eclipse pushed its path through western Asia on May 28, 585 B.C., passing over a battlefield where the armies of Lydia (a land in what is now western Turkey) and Media (a land in what is now northern Iran) were getting ready to fight. The eclipse to them was a clear sign that the gods were angry, so they hastily patched up a treaty of peace and disbanded their armies.

Although I am a rationalist, I can't help but sigh for the simple faith of those times. Would that the nations of today would recognize the ongoing disasters of our generation as the displeasure of whatever gods there be and disband *their* armies.

7. Updating the Asteroids

I was on a morning TV talk show here in New York City not long ago and as I prepared for the taping a few days earlier, one of the young ladies in charge said to me, "Oh, it's so exciting, Dr. Asimov, that you knew all about what was going to happen today, so many years ago."

"I did?" I said, trying not to show my astonishment, for as a science fiction writer I'm *supposed* to know about the future.

"Yes," she said. "Mike Wallace interviewed you for the New York *Post* back in 1957 and asked you to predict something about the future, and you said that in the future, energy would become so hard to get that the government would tell you how warm to keep your house."

"You're kidding!" I said with blank surprise, completely forgetting that I was a science fiction writer who was *supposed* to know the future.

"No, I'm not," she said and showed me a yellowed clipping and, my goodness, that was exactly what I said back in 1957.

I recovered with some difficulty and said, with as near an approach to benign omniscience as I could manage, "Ah, yes! And is there anything else you would like to know?"

Would that my batting average were always so accurate, but it isn't. And would that I could always foresee new scientific discoveries but I can't.

Therefore, every once in a while I feel I ought to update and amplify some earlier essay in this series. The particular essay I have in mind at the moment is "The Rocks of Damocles," which appeared in *From Earth to Heaven* (Doubleday, 1966).

In that article I mentioned seven "Earth-grazers"—seven asteroids that could approach Earth more closely than any planet did. They were Albert, Eros, Amor, Apollo, Icarus, Adonis, and Hermes. Now, however, helped along by statistics in a delightful article by Brian G. Marsden in the September 1973 *Sky and Telescope*, I would like to go over the subject again more thoroughly and with a different emphasis.

In any book on astronomy, including those I myself have written, you will find the statement that the asteroid belt is located between the orbits of Mars and Jupiter. It is, in the sense that between those orbits is present the largest concentration of asteroids.—But can there be asteroids with orbits carrying them beyond those limits?

Let's begin by considering the orbits of Mars and Jupiter. The orbits of both planets are elliptical, of course, and in both cases, the planet has a nearest approach to the Sun (perihelion) and a farthest approach (aphelion) that are distinctly different.

In the case of Mars, the perihelion distance, in millions of kilometers, is 208 and the aphelion distance is 250. For Jupiter, the two figures are 740 and 815.

If we interpret this business of "between the orbits of Mars and Jupiter" at its narrowest, we would expect each asteroid never to recede as far as 740 million kilometers from the Sun (Jupiter's perihelion) or to approach closer than 250 million kilometers to the Sun (Mars' aphelion).

This turned out to be quite true of the first four asteroids discovered: Ceres, Pallas, Juno, and Vesta. The perihelion in millions of kilometers for these four asteroids are, respectively, 380, 320, 300, and 330. The aphelions of the four asteroids are, respectively, 450, 510, 500, and 385. All fall neatly between the orbits of Mars and Jupiter.

Once these four asteroids were discovered, then, the notion that asteroids fall into the Mars-Jupiter gap was firmly established.

Yet suppose there are exceptions. These could be of two types. An asteroid might recede beyond Jupiter's orbit, or it might approach within Mars' orbit. (Or, of course, it might do both.)

The possibility of receding beyond Jupiter's orbit is somehow less disturbing psychologically. In the first place, the farther an asteroid, the more difficult it is to see, and an asteroid receding beyond Jupiter's orbit at aphelion might very likely be unusually far away at perihelion, too, and might be difficult to spot. Then, too, distance lends uninterest and a beyond-Jupiter asteroid would have nothing but statistics to recommend it.

On the other hand, an asteroid with an orbit that brought it within Mars' orbit at any point would be closer to us than asteroids would ordinarily be and would be correspondingly easier to detect. Furthermore, a nearer-than-Mars asteroid would be approaching Earth and for that very reason would be of devouring interest, whether because of the scientific studies it might make possible or because of the grisly chance that it might someday be on a collision course with ourselves.

To be sure, space is enormous even inside the comparatively constricted area of Mars' orbit (constricted as compared with the vast areas through which the outer planets course) and Earth is tiny by comparison. Then, too, the asteroids and Earth have orbits that tilt to different degrees, so that there is still more space available in the third dimension for the hoped-for miss. Nevertheless, any orbit that brings a maverick asteroid close to ourselves is bound to give rise to a certain concern.

Let us, therefore, concentrate on the possibility of asteroids moving within Mars' orbit and penetrating the inner solar system.

For seventy-two years after the discovery of the first asteroid, Ceres, everything went in orderly fashion. In that period, 131 asteroids were discovered and every one of them, without exception, had a perihelion distance of more than 250 million kilometers and therefore lay beyond Mars' farthest distance.

But then came June 14, 1873, when the Canadian-American astronomer James Craig Watson discovered asteroid 132. He named it Aethra (the mother of the Athenian hero Theseus in the Greek myths), and when he calculated its orbit, he found that the perihelion was less than 250 million kilometers. Aethra, at perihelion, approached the Sun more closely than Mars' distance at aphelion.

Aethra wasn't much closer than 250 million kilometers. Mars, at perihelion, was closer to the Sun than Aethra ever was. Still, Aethra was at times closer to the Sun than Mars at times was, and that was notable.

Three hundred more asteroids were discovered over the next quarter century, and then came August 13, 1898, when Gustav Witt discovered asteroid 433.

Its orbit was a real blockbuster, for its perihelion was at only 170 million kilometers from the Sun and that was *far* within Mars' orbit. The rule about asteroids orbiting between Mars and Jupiter was broken with a vengeance.

In fact, for the first time in the ninety-seven years since the first asteroid was discovered, it became appropriate to talk about Earth's orbit in connection with asteroids. Earth's perihelion is at 147 million kilometers from the Sun, and its aphelion is at 153 million kilometers.

If the orbits of Earth and of asteroid 433 are plotted in three dimensions, it turns out that the closest approach of the two is 23 million kilometers. In order for the two bodies to approach one another at this distance, each has to be in the particular point in its orbit that is nearest the other. More often than not, when one object is at the correct point, the other is far from it. For that reason, a reasonably close approach comes about only at long intervals.

Still, every once in a while, asteroid 433 does make a pretty close approach, and when it does, it is closer to us than any of the large planets of the solar system. Mars never comes closer to us than 55 million kilometers and Venus, our closest planetary neighbor, is never closer than 40 million kilometers.

Anything that can approach Earth at a distance of less than 40 million kilometers is therefore called (perhaps a bit overdramatically) an "Earth-grazer," and asteroid 433 was the first object of this sort to be discovered.

Witt called his new asteroid Eros, after the young god of love, who was the son, in the Greek myths, of Ares (Mars) and Aphrodite (Venus). This was a most appropriate name for an object which moved through space between the orbits of Mars and Venus (at least part of the time). It set the fashion of giving masculine names to all asteroids whose orbits lay, at least in part, within that of Mars or beyond that of Jupiter, or both.

For thirteen years and nearly three hundred more asteroids, Eros remained in lonely splendor as the only Earth-grazer known. Then in 1911 asteroid 719 was discovered and was found to penetrate within Mars' orbit. It was named Albert, and its perihelion distance was larger than that of Eros so that it can approach Earth no closer than 32 million kilometers.

Albert was lost after its discovery, a not unusual event. Orbits are not always calculated with sufficient precision during the short time asteroids can be observed after their initial (usually accidental) sighting, at a moment when they may already be almost out of sight. Then, too, their orbits can be perturbed by the various planets and changed slightly. Between lack of precision and possible orbital change, they may not come back in the place and at the time predicted; and they are usually faint enough to be missed *unless* you know in advance exactly when and where they will reappear.

In subsequent years, asteroids 887 (Alinda) and 1036 (Ganymed, with the final "e" missing to distinguish it from Jupiter's large satellite) were found to have orbits very like that of Albert, so the total number of Earth-grazers was raised to four.

In 1931, thirty-three years after its discovery, Eros still held the record for close approach to Earth. In that year Eros made a comparatively close approach—one of 27 million kilometers—and it was then the object of a large international project, since by measuring its parallax accurately, the scale of the solar system could be (and was) determined with record precision.

But then, on March 12, 1932, the record was broken when a Belgian astronomer, Eugène Delporte, discovered asteroid 1221 and found that its orbit gave it a perihelion distance of 162 million kilometers from the Sun, 8 million kilometers closer than Eros ever came. Moreover, it could approach within 16 million kilometers of Earth, which was considerably less than Eros' closest possible approach of 23 million. Delporte gave asteroid 1221 the name of Amor, which is the Latin equivalent of the Greek Eros.

In my 1966 article I mentioned Amor and said that it had a diameter of 16 kilometers.* Apparently, that's wrong. According to Marsden, Amor is the faintest of all the asteroids which have been honored with an official number and it must be somewhat less than 1 kilometer in diameter.

But if Amor, at the time of its discovery, held the record for close approach to the Sun, it held that record for only six weeks. On April 24, 1932, the German astronomer Karl Reinmuth discovered an asteroid he named Apollo. It was appropriate to name it after the Greek god of the Sun, for when the orbit was worked out, it turned out that this sixth Earth-grazer did not merely penetrate within Mars' orbit as the other five did. It went past the orbit of

---

* Actually, I said "10 miles" but I'm trying not to use naughty words like "miles" and "pounds" any more.

Earth, too, and even that of Venus, and ended up on July 7, 1932, at a perihelion point only 97 million kilometers from the Sun. (Venus' nearly circular orbit keeps that planet 109 million kilometers from the Sun.)

On the night of May 15, 1932, when it happened to be photographed at Yerkes Observatory, Apollo was only 11 million kilometers from Earth and had come closer than any other known deep-space object, except for our own Moon, of course. (I say "deep-space" because meteors come closer still, but they are seen only after they are within our atmosphere.)

Apollo is called 1932 HA, meaning that it was the first asteroid (A—first letter of the alphabet) to be discovered in the eighth half-month (H—eighth letter of the alphabet) of 1932. Its orbit, as worked out, seemed too uncertain to make it reasonable to suppose it would be seen again, so it was not given a number. And, indeed, it was promptly lost. However, Marsden and his co-workers have located it again in 1973 by making a concerted attempt to look where it might be. (And it was.)

The discovery of Apollo meant there was now a new class of objects, "Apollo-objects," which were asteroids that approached more closely to the Sun than the Earth does.

Apollo was the first of the Apollo-objects, but as you might guess, it wasn't the last. In February 1936 Delporte, who had detected Amor four years earlier, detected another object, perhaps just as small, which he named Adonis, after a well-known love of Aphrodite (Venus). Its official name is 1936 CA and it, too, is less than 1 kilometer in diameter (whereas Apollo is slightly larger, perhaps 1.5 kilometers).

Adonis, despite its small size, was brighter than Amor (and streaked more rapidly through the skies) because it was unusually close to the Earth at the time of its discovery. The calculation of its orbit showed that it had passed 2.4 million kilometers from the Earth a few days before it was detected—just ten times farther than the Moon.

What's more, the astonishing record set by Apollo for perihelion distance was broken, for at its closest approach, Adonis is only 66 million kilometers from the Sun, only two thirds the distance of either Apollo or the planet Venus.

Indeed, with Adonis, we can begin talking about the planet Mercury, the closest of all planets to the Sun. Mercury's orbit is more elliptical than that of the other three planets of the inner solar system. At its perihelion it is 46.5 million kilometers from the Sun, but at aphelion it is 70.5 million kilometers from the Sun. This means that Adonis is, at times, closer to the Sun that Mercury is.

Adonis' perihelion record remained for a number of years, but its record close approach to Earth did not. In November 1937 Reinmuth (the discoverer of Apollo) detected a third Apollo-object and named it Hermes. Properly called 1937 UB, Hermes moved across the sky with extraordinary quickness and was gone almost before a stab could be made at calculating its orbit. At its perihelion it was 93 million kilometers from the Sun, a little closer than Apollo.

It turned out, though, that Hermes passed within 0.8 million kilometers of the Earth and, if its orbit was calculated correctly, it was possible for it to miss us by a mere 0.31 million kilometers. It would then pass by us more closely than our own Moon does.

Hermes set a close-approach record that still exists (and that no one is eager to have broken). The only objects that have come closer to Earth than 0.31 million kilometers, that we know of, have penetrated Earth's atmosphere.

Of course, 0.31 million kilometers is twenty-four times the Earth's diameter and is a decided miss, but asteroids *can* be perturbed into a new orbit and we've lost track of Hermes. It has never been spotted since that 1937 fly-by and the orbital calculation isn't good enough to make us certain when it will be close enough to see again. It may be spotted again someday but only by accident, and meanwhile it's flying around in space with the potential of coming awfully close.

If it's any consolation to us (and why should it be?), other planets are also targets. Apollo passed by Venus at a distance of only 1.4 million kilometers in 1950.

In 1948 an Apollo-object was discovered by the German-American astronomer Walter Baade, which turned out, in some ways, to be the most unusual asteroid of all, one which he named "Icarus." Its orbit was well enough calculated to make it seem worth an asteroid number—1566. Its period of revolution is only 1.12 years, as compared with the typical asteroidal period of 4.6 years for those that are well-behaved enough to remain within the Mars-Jupiter orbital gap.

This means that, on the whole, Icarus' orbit is very little larger than Earth's. Icarus' orbit, however, is much more eccentric than

Earth's so that at its aphelion it is farther from the Sun than the Earth is, and at perihelion it is closer.

It is the perihelion that is remarkable, for at that point, Icarus is only 28.5 million kilometers from the Sun, less than half the distance of Adonis, the previous record holder, and only about 60 per cent of the closest approach Mercury ever makes to the Sun.

It is appropriate that this asteroid be named for the character in the Greek myths who flew too near the Sun (which melted the wax holding the feathers onto his artificial wings) and fell into the sea to his death. Nothing we know of ever approaches the Sun more closely than Icarus, except for an occasional comet, like that observed in 1843.

(If you're curious, the comet of 1843, as it swung about the Sun approached to within 0.825 million kilometers of the Sun's center, or within 0.13 million kilometers of its surface. That it survived was owing to the fact that it was moving at the kind of speed that enabled it to swing around the Sun in a little over an hour behind the insulation of the gas and dust that arose out of its evaporating ices— and then to get the devil away from there.)

Icarus' perihelion value has kept the record for the quarter century since it has been discovered. Nothing else has even come near that value. It is no record holder as far as its approach to Earth is concerned. The closest it can come to Earth is 6.4 million kilometers.

In 1951 the French-American astronomer Rudolph L. B. Minkowski, along with E. A. Wilson, discovered Geographos, which received the asteroid number 1620 and had the unusually large diameter for an Apollo-object of about 3 kilometers. Its perihelion distance is about 124 million kilometers and its closest approach to Earth is about 10 million kilometers.

One other numbered Apollo-object should be mentioned. It was discovered in 1948, but wasn't given a number till it was reobserved by Samuel Herrick of the University of California in 1964. It then received the number 1685 and the name Toro. Toro is even larger than Geographos, about 5 kilometers across. Its perihelion distance is 115 million kilometers and it never comes closer to Earth than 15 million kilometers.

It turned out, however, that Toro's orbit is locked in with that of the Earth. That is, Toro moves out from the Sun and in toward the Sun in an orbit that keeps step with that of Earth, Toro moving around the Sun five times every time Earth moves around eight times.

It is as though Earth and Moon are engaged in a close-held waltz in the center of the ballroom while far at one end, a third person is engaged in an intricate minuet of its own, one that carefully stalks, from a distance, the waltzing couple. Every so often Toro shifts from an Earth-lock to a Venus-lock, and then back to Earth. Its period with Earth is considerably longer than its period with Venus.

Aside from the Apollo-objects mentioned above, eleven other Apollo-objects have been discovered as of mid-1973. Two of them, 1971 FA and 1971 UA, approach the Sun more closely at perihelion than any other Apollo-objects except for Icarus and Adonis. Another one, 1972 XA, is the largest Apollo-object yet detected, with a diameter of perhaps more than 6 kilometers, while 1973 NA has an orbital inclination of 67°, which is something only seen in a cometary orbit.

Maybe, for that matter, some of the Apollo-objects *are* comets—dead ones—comets, perhaps, that have lost all their volatile materials long ago and that happened to have a compact rocky nucleus of from 1 to 6 kilometers in diameter still moving along in the old orbit.

Encke's Comet, for instance, which has a perihelion distance of 51 million kilometers (second only to Icarus among the Apollo-objects), has very little haze about itself and also has an apparently compact core. Once all its volatiles are gone, what can we possibly call it but an Apollo-object?

In Table 6 there is a list of the known Apollo-objects, in which I have included for comparison the five innermost planets and Encke's Comet.

As you see from the table, only one Apollo-object, Icarus, comes closer to the Sun than Mercury ever does. Six others, in addition, including Adonis, Hermes and Apollo (and Encke's Comet as a seventh, if you want to count it) come closer to the Sun than Venus ever does. Ten others, in addition, including Toro and Geographos, come closer than Earth ever does.

However, don't let that perihelion value fool you completely. None of the Apollo-objects *stays* close. Every single one of the Apollo-objects recedes, on each revolution, beyond the orbits of Mercury, Venus, and Earth.

TABLE 6 — THE APOLLO-OBJECTS

Planet	Apollo-object	Perihelion (millions of kilometers)	Aphelion (millions of kilometers)
	Icarus	28.5	300
Mercury		46.5	70.5
	(Encke's Comet)	51	615
	Adonis	66	495
	1971 FA	84	360
	1971 UA	87	240
	Hermes	93	405
	1973 EA	93	435
	Apollo	97	345
Venus		109	109
	Toro	116	300
	P-L 6743	124	360
	Geographos	125	255
	1947 XC	125	555
	1959 LM	125	285
	1950 DA	127	375
	1972 XA	132	435
	1973 NA	133	555
	1948 EA	135	540
	P-L 6344	142	630
Earth		148	152
Mars		208	250
Jupiter		740	815

The Apollo-object 1971 UA has the smallest aphelion, 240 million kilometers, so that it alone never recedes as far as Mars does. All the other Apollo-objects, however, including even Icarus, recede well beyond Mars' orbits and all are full-fledged members of the asteroid belt, at least in the distant portion of their orbits.

No known Apollo-object recedes as far as Jupiter's orbit, however. Even Encke's Comet does not. Encke's Comet is, at all times, well within Jupiter's orbit, and it is the only known comet of which this can be said. All others, without exception have larger orbits and recede beyond Jupiter.

But suppose we consider the average distance of any object from the Sun as roughly equal to halfway between aphelion and perihelion. In that case, if we line up the Apollo-objects in terms of average distance, we get Table 7.

What it amounts to is that no Apollo-object, not even Icarus, is as close to the Sun, on the average, as even *Earth* is. Only 1971 UA and Icarus approach Earth's mark and vie for the shortest known period of revolution for any asteroid—a little over 400 days for each.

TABLE 7 — THE APOLLO-OBJECTS—AVERAGE

Planet	Apollo-object	Average distance (millions of kilometers)
Mercury		58
Venus		109
Earth		150
	1971 UA	163
	Icarus	164
	Geographos	190
	1959 LM	205
	Toro	208
	Apollo	221
	1971 FA	222
Mars		229
	P-L 6743	242
	Hermes	249
	1950 DA	251
	1973 EA	264
	Adonis	280
	1972 XA	283
	(Encke's Comet)	333
	1948 EA	338
	1947 XC	340
	1973 NA	344
	P-L 6344	386
Jupiter		778

Five more are slightly less distant from the Sun, on the average, than Mars is, while ten (plus Encke's Comet) are farther from the Sun than Mars is. Not one of them is even half as far from the Sun, on the average, as Jupiter is, though P-L 6344 almost makes it.

Mercury, Venus, and even the Earth and the Moon retain their pride of place as being nearer the Sun on a day-in, day-out basis than any other known object in the solar system, including even the comets.

And how many more Apollo-objects are there yet undiscovered? The American astronomer Fred Whipple suspects that there may be at least a hundred larger than 1.5 kilometers in diameter, and it follows that there must be thousands of additional ones that are less than 1.5 kilometers in diameter but large enough to do damage if they blunder into us (and perhaps tens of thousands of objects small enough to be no more than an annoyance when they strike).

Every meteoroid that hits our atmosphere and is large enough to make its way through it and strike the ground as a meteorite was an Apollo-object before the collision. Few are large enough to be catastrophic, though at least two twentieth-century strikes would have been, if kindly Chance had not seen fit to guide them to unpopulated Siberian areas.

The astronomer Ernst Öpik estimates that an Apollo-object should travel in its orbit for an average of 100 million years before colliding with Earth. If we suppose that there are two thousand Apollo-objects large enough to wipe out a city if they strike, then the average interval of time between strikes of any of these is only 50,000 years.

The great Meteor Crater in Arizona may have been formed by one of the smaller, but yet sizable, Apollo-objects, and that may have been formed about 50,000 years ago.

Maybe we're due.†

† Shortly after this was written, it was reported that on August 10, 1972, a small Apollo-object passed through the upper regions of the atmosphere and out again in a glancing blow. When it was over southern Montana it passed within 50 kilometers of Earth's surface. It was only 13 meters across, but if it had struck a populated area, it would undoubtedly have wiped out several city blocks and done damage for a considerable distance about.

C—LIFE

8. The Inevitability of Life

I belong to the Gilbert & Sullivan Society here in New York, and I attend every meeting I can and sing with gusto. Naturally, the club makes much of me, because they're a bunch of very goodhearted people. At this month's meeting, a gentleman brought over a very nice young lady, and said, "This is my daughter. She wrote a report on you for school."

"Ah," I said, smiling genially, "and how old are you, young lady?"

"Twelve," she said, shyly.

I shook my head reminiscently, "I was once twelve," I said. "I'm a little older than that, now."

"Not so little," said the young lady, not at all shyly. "You were —— last January 2."*

She then sat down next to me and we all sang from *The Gondoliers*. She refused my generous offer to share the book with her and proceeded to sing all the songs perfectly, missing no words and missing no notes—even in that complicated patter song that begins "Rising early in the morning."

I have rarely spent an evening in which my spirit was so completely broken.

Oh well, I have my outlets. A lesser man might merely say, "That's life!" but I can seek forgetfulness not only by referring to the vagaries of life, but by going into the subject of life in detail. As follows—

In the course of the nineteenth century, scientists began to accept the concept of biological evolution and to dismiss the possibility that

---

* I omit the actual number she used because I don't think it would interest you, and I don't want to bore you.

life had been created, ready-made in its present complexity, by some supernatural agency.

This had its uncomfortable aspect, for it raised the question as to how the extraordinary phenomenon of life, with all its intricacy and versatility, could possibly have come to be "by accident." From what simple chemical starting point could it have stemmed, and by what processes could it have reached that delicate stage we call "living"? It seemed too much to ask of chance, and if we look about the Earth today, we certainly see no spontaneous chemical changes taking place that seem to be moving in the direction of the development of life unless life is involved to begin with.

But then, we can't expect life to develop on Earth today. In the first place, Earth's chemistry is not what it was at the time life must have originated, having been changed massively since then by life itself. Then, too, life forms already existing would quickly eat, or otherwise alter, any substances produced today that are, so to speak, part way toward life.

To consider the manner in which life originated, we must try to determine what conditions were like on Earth before life had formed and consider only those changes that could take place in the absence of any form of life.

Since astronomical evidence points to the fact that some 90 per cent of the atoms in the Universe are hydrogen, it would seem that planetary atmospheres, to begin with, must consist of hydrogen, plus combinations of hydrogen atoms with atoms of other fairly common elements. Thus, Jupiter's atmosphere consists chiefly of hydrogen molecules (H_2), plus minor quantities of the carbon-hydrogen combination, CH_4, or methane, and the nitrogen-hydrogen combination, NH_3, or ammonia. The oxygen-hydrogen combination, H_2O, or water, is undoubtedly also present but not in the upper reaches of Jupiter's atmosphere, which is all we can observe.

The American chemist Harold C. Urey, who turned to the consideration of such problems after World War II, suggested that the origin of life be considered in connection with an atmosphere something like that of Jupiter.

One of his graduate students, Stanley L. Miller, tried, in 1952, to duplicate primordial conditions on Earth. He began with a closed and sterile mixture of water, ammonia, methane, and hydrogen, representing a small and simple version of Earth's primordial atmosphere and ocean. He then used an electric discharge as an energy source, representing a tiny version of the Sun.

He circulated the mixture past the discharge for a week and then analyzed it. The originally colorless mixture had turned pink and one sixth of the methane with which he had started had gone into the "abiotic formation" (formation without the intervention of living organisms) of more complex molecules. Among those molecules were glycine and alanine, the two simplest amino acids of those that form the basis of proteins, which are, in turn, one of the two classes of compounds most characteristic of living tissue.

For twenty years similar experiments were conducted, with variations in starting materials and in energy sources. Invariably, more complicated molecules, sometimes identical with those in living tissue, sometimes related to them, were formed. An amazing variety of key molecules of living tissue were formed abiotically in this manner.

This was done in small volumes over very short periods of time. What could be done in an entire ocean over a period of millions of years? And all the changes produced by the chance collisions of molecules and the chance absorptions of energy seemed to move in the direction of life as we know it now. There seemed no important changes that pointed definitely in some different chemical direction.

It might be concluded, then, that life originated through the most likely chemical changes undergone by the most common molecules present in the atmosphere and ocean of the primordial Earth. From that viewpoint, life was by no means an excessively rare and miraculous cosmic accident that seemed to require some supernatural agency for its beginnings.

Rather, it was the inevitable result of high-probability chemical reactions, and any planet in the Universe possessing a chemistry vaguely like that of Earth and bathed in the light of a star vaguely like that of our Sun should develop life as Earth did. It is partly on the basis of such reasoning that many astronomers now believe that there are many billions of life-bearing planets in the Universe.

But are we right in assuming that Miller and other experimenters produced amino acids and other molecules abiotically? Wasn't life involved in the form of the human experimenter?

The experimenter chooses the starting conditions and dictates the course of the experiment. It may be taken for granted that he reproduces what he honestly believes to be a close version of what may likely be the truth of the situation, and yet the experimenter knows the answer he can't help wanting—at least, he knows the chemical basis of earthly life as it is today. May it not be, then, that he is unconsciously influenced by that and directs the experiment in such a way as to get the answer he expects?

It would be much better if some evidence could be found concerning the origins of life that did not involve human-directed experimentation. —But, short of a time machine that would take us back to Earth's origins, how?

It might be possible to consider another world that is something like the Earth in chemical composition and temperature, but sufficiently different, perhaps, to abort the development of life and level it off at some stage far less complex than that existing on Earth. Such a world would offer an environment closer to that of the primordial Earth than the present-day Earth does and might give us useful information as to the very early stages of biological evolution— or even of the chemical evolution that preceded life.

In the early 1950s, when the modern laboratory investigations of the origin of life began, three worlds might have been considered to fall into the "sufficiently Earthlike" category: Venus, Moon, and Mars. Of these three, Venus was crossed off in 1962 when the Mariner II space probe showed, with final certainty, that Venus' surface was far too hot for any form of chemical evolution applicable to earthly life.

In 1969 the Moon was crossed off. The first sample of rocks brought back from our satellite showed that it was disappointingly low in those elements most involved in life. In particular, it seemed completely water-free, and no useful traces of Earthlike chemical evolution were to be found in its soil.

That leaves Mars, which is known to have volatile material—a thin carbon dioxide atmosphere, with possibly some nitrogen, and also icecaps that contain, in all probability, both carbon dioxide and water in frozen state. Its temperature is lower than that of Earth but not too low, and it is conceivable that simple life forms may exist on it—or, failing that, advanced stages of preliving chemical evolution.

It is the thought that Mars might be an untouched model of a primordial Earth, a thought that is one of the most exciting driving forces behind the attempts to send life-detecting systems (or, better yet, human beings) to a soft-landing on Mars.

Yet this does not exhaust those worlds in the neighborhood of Earth that might be useful with respect to the origin of life. What of the meteorites—excessively tiny worlds, but worlds none the less. We needn't go out into space to study them; they have already come to us.

About 1,700 meteorites have been studied; thirty-five of them weighing over a ton apiece; but almost all of them are either nickel-iron or stony in chemical composition and in either case show no signs of useful information concerning our problem.

There remains, however, a third, and very rare, type of meteorite —it is black and easily crumbled. A meteorite like this contains small stony inclusions that lend the object a grainy appearance, and these inclusions are called "chondrules," from a Greek word for "grains." Meteorites of this type are "chondrites," and because they possess carbon-containing compounds, they are called "carbonaceous chondrites." Less than two dozen such meteorites are known.

The first carbonaceous chondrite known to have fallen dropped near Alais, France, in 1806. In 1834 the Swedish chemist Jöns J. Berzelius studied it and wondered if its carbonaceous matter had once been part of living organisms. In the early 1960s there were reports of microscopic objects in a carbonaceous chondrite that had fallen near Orgueil, France, in 1864, objects that actually looked once-living. That turned out to be a false alarm, however. The meteorite, lying about for a hundred years, had been contaminated by earthly pollen and so on.

What was needed was a freshly fallen carbonaceous chondrite.

One such fell near Murray, Kentucky, in 1950, and another exploded over the town of Murchison in Australia on September 28, 1969. In the latter case, some 82.7 kilograms of fragments were recovered.

The Murchison and Murray meteorites were examined closely by NASA scientists at Ames Research Center in Iowa under the leadership of Keith Kvenvolden, and such modern techniques as gas chromatography and mass spectroscopy were used to first separate and then identify the components of the carbonaceous content of the meteorites.

In 1970 and 1971 it was determined that most of the isolated components were amino acids. Six of them were varieties that occur frequently in earthly proteins: valine, alanine, glycine, proline, aspartic acid, and glutamic acid. The remaining twelve of the eighteen detected were related, but were present in earthly living tissue in only small quantities if at all.

Similar results were obtained for the Murray meteorite. Agreement between the two meteorites that fell on opposite sides of the world nineteen years apart were impressive.

Toward the end of 1973 fatty acids were also detected in the Murchison meteorite. These differ from amino acids in having longer chains of carbon and hydrogen atoms and in lacking nitrogen atoms. They are the building blocks of the fat found in living tissue, and some seventeen different fatty acids were identified.

How did such complicated organic molecules happen to be found in meteorites? By far the most glamorous explanation would be that meteorites are the remnants of an exploded but once-life-bearing planet. Such a planet might have a nickel-iron core and a stony mantle, as the Earth has, and these would give rise to the two kinds of common meteorites. Perhaps the outermost crust of the planet bore life forms, and the carbonaceous chondrites are pieces of that crust still containing traces of materials that had been formed by ancient life.

Apparently, this is not likely. There are ways of testing whether the compounds discovered in meteorites are likely to have originated in living things.

Amino acids (all except glycine, which is the simplest) come in two varieties, one of which is the mirror image of the other. These are labeled as L and D. The two varieties are identical chemically, so when chemists prepare amino acids by processes that don't involve any chemicals obtained from tissues, equal quantities of L and D are always formed.

Protein molecules cannot, however, be built up of L and D amino acids in random order; the geometry leaves no room for the atom combinations. There *is* room if all the amino acids are L—or if all are D.

On Earth, living tissue got started somehow, perhaps by sheer chance, with L amino acids. The chemical reactions within tissue are mediated by enzymes built up of L amino acids only, and only L amino acids are produced. This is true for all earthly organisms from viruses to whales.

If the amino acids from the meteorites were all L or all D, we would strongly suspect that life processes similar to our own were involved in their production. In actual fact, however, L and D forms are found in equal quantities in the carbonaceous chondrites, and this means that they originated by processes that did not involve enzymes, and therefore did not involve life as we know it.

Similarly, the fatty acids formed in living tissue are built up by the addition to each other of varying numbers of two-carbon-atom compounds. As a result, almost all fatty acids in living tissue have an even number of carbon atoms. Fatty acids with odd numbers are not characteristic of our sort of life, but in chemical reactions that don't involve life they are as likely to be produced as the even variety. In the Murchison meteorite, there are roughly equal quantities of odd-number and even-number fatty acids.

The Orgueil meteorite, although contaminated, seems to contain abiotic material, too. Among these, it is reported, are purines and pyrimidines which are among the building blocks of nucleic acids, that class of compounds which, in addition to proteins, are particularly characteristic of life on Earth today.

The compounds in the carbonaceous chondrites are not life but have been formed in the *direction* of our kind of life—and human experimenters have had nothing to do with their formation. On the whole, then, meteoritic studies tend to support laboratory experiments and make it appear all the more likely that life is a natural, a normal, even an inevitable phenomenon. Atoms apparently tend to come together to form compounds in the direction of our kind of life whenever they have the least chance to do so.

Yet how much reliance can we place on carbonaceous chondrites? They are rare objects and we don't know their history. Perhaps they are subject to conditions so unusual that it is unfair to use them to support the notion of life as of frequent occurrence.

What else can we study? Outside of our own solar system, the most easily observed objects are the stars, all of which are incandescently hot and therefore of no direct use to us as guides to the chemical evolution of life.

Still, there must be cold portions of the Universe, too. There are uncounted billions of planets circling other stars, but none of them are directly detectable by us. The existence of a very few can be deduced from the wavering motions of the stars they circle, but that only tells us the planetary mass and nothing more.

The only detectable examples of cold matter outside our solar system are the thin gas and dust in interstellar space. Though that must surely seem as unlikely a place for chemical evolution as the stars themselves, let's look at it.

The interstellar medium was detected about the turn of the century because certain wavelengths of light from distant stars were absorbed by the occasional atoms that drift about in the vastness of space. By the 1930s it was recognized that the interstellar medium was made up largely of hydrogen atoms, with admixtures of small amounts of other atoms: helium, carbon, nitrogen, oxygen, and so on. Probably some of every type of atom is present, but with hydrogen overwhelmingly predominant.

The density of the interstellar matter is so low that it is easy to imagine it as consisting almost entirely of single atoms and nothing else. After all, in order for two atoms to combine to form a molecule, they must first collide, and the various atoms are so widely spread apart in interstellar space that random motions will bring about collisions only after excessively long periods. This means that two-atom combinations may well exist there but in such tiny concentrations as to be indetectable, and concerning three-atom combinations there would be no need to talk at all.

There are also dust particles present in outer space. They are known to exist, since dark clouds that hide the starlight behind them are common in the region of the Milky Way. Individual atoms absorb little light, while dust particles absorb much, so the dark clouds must contain a considerable quantity of dust. (The chemical nature of this dust and how it formed are as yet matters of dispute, however.)

To detect two-atom combinations in interstellar space would be difficult indeed, in view of their inevitably low concentration. It could be done with ordinary telescopes only if such molecules happened to be between us and a particularly bright star, if they happened to occur in great enough concentration to absorb a detectable quantity of light, and if each one did so at wavelengths sufficiently characteristic of itself to allow itself to be identified.

In 1937 just such requirements were met and a carbon-hydrogen combination (CH, or methylidyne radical) and a carbon-nitrogen combination (CN, or cyanogen radical) were detected.

For the first time, interstellar molecules were found to exist. CH and CN were the kind of combinations that could be formed and maintained only in very low-density material, for such atom combinations are very active and would combine with other atoms instantly, if other atoms were easily available. It is because such other atoms are available in quantity on Earth that CH and CN do not exist naturally, as such, on our planet.

These molecules, however, could only be seen in connection with very bright (and therefore very rare) stars and could not be observed elsewhere, if indeed they existed elsewhere. They were little more than curiosities, and optical telescopes discovered no other examples of interstellar molecules.

After World War II, however, radio astronomy became increasingly important. Interstellar atoms could emit or absorb characteristic radio waves—something that requires far less energy than would be needed to emit or absorb visible light waves. The emission or absorption of such radio waves could be detected easily and would be characteristic enough to identify the molecules involved. In 1951, for instance, the characteristic radio-wave emission by hydrogen atoms was detected, and interstellar hydrogen was for the first time directly detected, instead of being merely deduced.

This was ordinary hydrogen, or hydrogen-1, its nucleus a single proton. In 1966 the slightly different radio-wave radiation of hydrogen-2, or deuterium (with a nucleus made up of a proton and a neutron), was also detected.

Meanwhile, though, it was understood that next to hydrogen, helium and oxygen were the most common atoms in the Universe. Helium didn't form compounds, but oxygen did. Should there not be oxygen-hydrogen combinations (OH, or hydroxyl radical) in space? Hydroxyl should emit radio waves in four particular wavelengths, and two of these were detected for the first time in 1963.

Even as late as the beginning of 1968, only three different atom combinations had been detected in outer space: CH, CN, and OH.

Each of these were two-atom combinations that seemed to have arisen from the chance, but very rare, collisions of individual atoms. (All three are atom combinations that are very common in living tissue, but that just means that the types of atoms that are common in tissue are also common in space.)

No one expected that the far less probable combination of *three* atoms would build up to detectable levels, and yet in 1968 observers at the University of California located radio-wave emissions from interstellar space that were characteristic of molecules of water and of ammonia. Water has a three-atom molecule (H_2O) and ammonia, a four-atom molecule (NH_3).

This was utterly astonishing and 1968 witnessed the birth of what we now call "astrochemistry."

In fact, once compounds of more than two atoms were detected, the list grew rapidly longer. In 1969 a four-atom combination involving the carbon atom was discovered. This was formaldehyde (H_2CO).

In 1970 the first five-atom combination was discovered (HCCCN, or cyanoacetylene) and the first six-atom combination (CH_3OH, or methyl alcohol) also. In 1971 the first seven-atom combination was discovered (CH_3CCH, or methylacetylene). At this time of writing, over two dozen molecules have been discovered in interstellar space.

Of the molecules now known to exist in space, those which are most widespread and common are hydroxyl (OH), formaldehyde (H_2CO), and carbon monoxide (CO). They are found throughout the Milky Way, while the others are found only among one or another of various dust clouds. (There may be about three thousand of these dust clouds in our Galaxy, each averaging twelve light-years across and containing a quantity of dust that may, on the average, total twenty times the mass of our Sun.)

In these dust clouds, atoms are more densely distributed and will collide more frequently. The incidence of molecule formation rises. Then, too, the dust particles may serve as nuclei on which atoms may condense and upon which interaction and combination can be hastened. Finally, the dust particles help block out the ultraviolet light from the stars—energetic light which might otherwise tend to break down the molecules as fast as they are formed.

With denser material to begin with, with the rate of formation increased, and with the rate of destruction decreased, it is not surprising, after all, that complex molecules are formed in large quantities in dust clouds.

The carbon atom seems to play a central role in molecules formation in the dust clouds, just as it plays a central role in living tissue. It does so, possibly, for the same reason in both cases—the versatile mananer in which it tends to form endlessly different combinations with other atoms.

In the interstellar dust clouds it may be that the great number of carbon monoxide molecules combine with the even greater number of atoms and molecules of hydrogen to form the large variety of carbon-containing molecules in space. (Three fourths of the different molecules that have so far been detected in interstellar space contain at least one carbon atom, even though the carbon atom is, at best, only the sixth most common of the atoms in space.) Such a reaction is called the "Fischer-Tropsch synthesis," and it has been known and used on Earth since 1923.

Life, then, would seem to be even more nearly inevitable than had begun to be suspected in the 1950s.

In the dust clouds of outer space, even before stars or planets have taken shape, molecules form that seem already to be on the road toward amino acids, fatty acids, purines, pyrimidines, and others of the basic building blocks of living tissue.

As these dust clouds condense here and there in space to form stars, the number and concentration of these molecules increase and still more complicated specimens are formed. Those parts of the clouds that actually form stars usually grow hot enough to break down all the molecules that have formed, but what of those parts of the clouds that remain cool and from which the planets are formed?

All the planets that are large enough, or cold enough, *begin* with compounds that give them a head start toward life. A planet as hot as Venus is large enough to contain some of these molecules—such as carbon dioxide—in vast quantities. An object as small as a meteorite is cold enough to retain small quantities of these space-born molecules, at least occasionally; and these seem to have built up to the complexity of fatty acids and amino acids.

Then, of course, if a planet is large enough and cool enough to retain a large quantity of a large variety of such molecules and is near enough to the Sun to receive a copious energy input—as Earth is—chemical evolution continues onward from its head start.

The origin of life? —On a planet like ours, there is simply no way to avoid life.

9. Look Long upon a Monkey

Considering that I work so hard at establishing my chosen persona as a man who is cheerfully self-appreciative, I am sometimes absurdly sensitive to the fact that every once in a while people who don't know me take the persona for myself.

I was interviewed recently by a newspaper reporter who was an exceedingly pleasant fellow but who clearly knew very little about me. I was curious enough, therefore, to ask why he had decided to interview me.

He explained without hesitation. "My boss asked me to interview you," he said. Then he smiled a little and added, "He has strong, ambivalent feelings about you."

I said, "You mean he likes my writing but thinks I am arrogant and conceited."

"Yes," he said, clearly surprised. "How did you know?"

"Lucky guess," I said, with a sigh.

You see, it's *not* arrogance and conceit; it's cheerful self-appreciation, and anyone who knows me has no trouble seeing the difference.

Of course, I could save myself this trouble by choosing a different persona, by practicing aw-shucks modesty and learning how to dig my toe into the ground and bring the pretty pink to my cheeks at the slightest word of praise.

But no, thanks. I write on just about every subject and for every age level, and once I begin to practice a charming diffidence, I will make myself doubt my own ability to do so, and that would be ruinous.

So I'll go right along the path I have chosen and endure the ambivalent feelings that come my way, for the sake of having the self-assurance to write my wide-ranging essays—like this one on evolution.

I suspect that if man* could only have been left out of it, there would never have been any trouble about accepting biological evolution.

Anyone can see, for instance, that some animals resemble each other closely. Who can deny that a dog and a wolf resemble each other in important ways? Or a tiger and a leopard? Or a lobster and a crab? Twenty-three centuries ago the Greek philosopher Aristotle lumped different types of species together and prepared a "ladder of life," by arranging those types from the simplest plants upward to the most complex animals, with (inevitably) man at the top.

Once this was done, we moderns could say, with the clear vision of hindsight, that it was inevitable that people should come to see that one type of species had changed into another; that the more complex had developed from the less complex; that, in short, there was not only a ladder of life but a system whereby life forms climbed that ladder.

Not so! Neither Aristotle nor those who came after him for more than two thousand years moved from the ladder of life as a static concept to one that was a dynamic and evolutionary one.

The various species, it was considered, were *permanent*. There might be families and hierarchies of species, but that was the way in which life was created from the beginning. Resemblances had existed from the beginning, it was maintained, and no species grew to resemble another more—or less—with the passage of time.

My feeling is that the insistence on this constancy of species arose, at least in part, out of the uncomfortable feeling that once change was allowed, man would lose his uniqueness and become "just another animal."

Once Christianity grew dominant in the Western world, views on the constancy of species became even more rigid. Not only did Genesis 1 clearly describe the creation of the various species of life as already differentiated and in their present form, but man was created differently from all the rest. "And God said, Let us make man in our image, after our likeness . . ." (Genesis 1:26).

---

* Anyone who reads these essays knows that I am a women's-libber, but I also have a love for the English language. I try to circumlocute "man" when I mean "human being" but the flow of sound suffers sometimes when I do. *Please* accept, in this article, "man" in the general, embracing "woman." (Yes, I know what I said.)

No other living thing was made in God's image and that placed an insuperable barrier between man and all other living things. Any view that led to the belief that the barriers between species generally were not leakproof tended to weaken that all-important barrier protecting man.

It would have been nice, of course, if all the other life forms on Earth were enormously different from man so that the insuperable barrier would be clearly reflected physically. Unfortunately, the Mediterranean world was acquainted, even in early times, with certain animals we now call "monkeys."

The various monkeys with which the ancients came in contact had faces that, in some cases, looked like those of shriveled little men. They had hands that clearly resembled human hands and they fingered things as human beings did and with a clearly lively curiosity. However, they had tails and that rather saved the day. The human being is so pronouncedly tailless and most of the animals we know are so pronouncedly tailed that that, in itself, would seem to be a symbol of that insuperable barrier between man and monkey.

There are, indeed, some animals without tails or with very short tails, such as frogs, guinea pigs, and bears, but these, even without tails, do not threaten man's status. And yet—

There is a reference to a monkey in the Bible, one for which the translaters used a special word. In discussing King Solomon's trading ventures, the Bible says (1 Kings 10:22), ". . . once in three years came the navy of Tharshish, bringing gold, and silver, ivory, and apes, and peacocks."

Tharshish is often identified as Tartessus, a city on the Spanish coast just west of the Strait of Gibraltar, a flourishing trading center in Solomon's time that was destroyed by the Carthaginians in 480 B.C. In northwestern Africa across from Tartessus, there existed then (and now) a type of monkey of the macaque group. It was this macaque that was called an "ape," and in later years, when northwestern Africa became part of "Barbary" to Europeans, it came to be called "Barbary ape."

The Barbary ape is tailless and therefore more resembles man than other monkeys do. Aristotle, in his ladder of life, placed the Barbary ape at the top of the monkey group, just under man. Galen, the Greek physician of about A.D. 200, dissected apes and showed the resemblance to man to be internal as well as external.

It was the resemblance to man that made the Barbary ape amusing to the ancients, and yet annoying as well. The Roman poet Ennius is quoted as saying, "The ape, vilest of beasts, how like to us!" Was the ape really the "vilest of beasts"? Objectively, of course not. It was its resemblance to man and its threat, therefore, to man's cherished uniqueness that made it vile.

In medieval times, when the uniqueness and supremacy of man had become a cherished dogma, the existence of the ape was even more annoying. They were equated with the Devil. The Devil, after all, was a fallen and distorted angel, and as man had been created in God's image, so the ape was created in the Devil's.

Yet no amount of explanation removed the unease. The English dramatist William Congreve wrote in 1695: "I could never look long upon a monkey, without very mortifying reflections." It is not so hard to guess that those "mortifying reflections" must have been to the effect that man might be described as a large and somewhat more intelligent ape.

Modern times had made matters worse by introducing the proud image-of-God European to animals, hitherto unknown, which resembled him even more closely than the Barbary ape did.

In 1641 a description was published of an animal brought from Africa and kept in the Netherlands in a menagerie belonging to the Prince of Orange. From the description it seems to have been a chimpanzee. There were also reports of a large manlike animal in Borneo, one we now call the orangutan.

The chimpanzee and the orangutan were called "apes" because, like the Barbary ape, they lacked tails. In later years, when it was recognized that the chimpanzee and orangutan resembled monkeys less and men more, they came to be known as "anthropoid" (manlike) apes.

In 1758 the Swedish naturalist Carolus Linnaeus made the first thoroughly systematic attempt to classify all species. He was a firm believer in the permanence of species and it did not concern him that some animal species closely resembled man—that was just the way they were created.

He therefore did not hesitate to lump the various species of apes and monkey together, *with man included as well*, and call that group

"Primates," from a Latin word for "first," since it included man. We still use the term.

The monkeys and apes, generally, Linnaeus put into one sub-group of Primates and called that subgroup "Simia," from the Latin word for "ape." For human beings, Linnaeus invented the subgroup "Homo," which is the Latin word for "man." Linnaeus used a double name for each species (called "binomial nomenclature," with the family name first, like Smith, John, and Smith, William), so human beings rejoiced in the name "Homo sapiens" (Man, wise). But Linnaeus placed another member in that group. Having read the description of the Bornean orangutan, he named it "Homo troglo-dytes" (Man, cave-dwelling).

"Orangutan" is from a Malay word meaning "man of the forest." The Malays, who were there on the spot, were more accurate in their description, for the orangutan is a forest dweller and not a cave dweller, but either way it cannot be considered near enough to man to warrant the Homo designation.

The French naturalist Georges de Buffon was the first, in the mid-dle 1700s, to describe the gibbons, which represent a third kind of anthropoid ape. The various gibbons are the smallest of the anthro-poids and the least like man. They are sometimes put to one side for that reason, the remaining anthropoids being called the "great apes."

As the classification of species grew more detailed, naturalists were more and more tempted to break down the barriers between them. Some species were so similar to other species that it was uncertain whether any boundary at all could be drawn between them. Besides, more and more animals showed signs of being caught in the middle of change, so to speak.

The horse, Buffon noted, had two "splints" on either side of its leg bones, which seemed to indicate that once there had been three lines of bones there and three hoofs to each leg.

Buffon argued that if hoofs and bones could degenerate, so might entire species. Perhaps God had created only certain species and that each of these had, to some extent, degenerated and formed additional species. If horses could lose some of their hoofs, why might not some of them have degenerated all the way to donkeys?

Since Buffon wished to speculate on what was, after all, the big news in man-centered natural history, he suggested that apes were degenerated men.

Buffon was the first to talk of the mutability of species. Here, however, he avoided the worst danger—that of suggesting that man-the-image-of-God had once been something else—but he did say that man could *become* something else. Even that was too much, for once the boundaries were made to leak in one direction, it would be hard to make it leakproof in the other. The pressure was placed on Buffon to recant, and recant he did.

The notion of the mutability of species did not die, however. A British physician, Erasmus Darwin, had the habit of writing long poems of indifferent quality in which he presented his ofttimes interesting scientific theories. In his last book, *Zoonomia*, published in 1796, he amplified Buffon's ideas and suggested that species underwent changes as a result of the direct influence upon them of the environment.

This notion was carried still further by the French naturalist Jean Baptiste de Lamarck, who, in 1809, published *Zoological Philosophy* and was the first scientist of note to advance a theory of evolution, a thoroughgoing description of the mechanisms by which an antelope, for instance, could conceivably change, little by little over the generations, into a giraffe. (Both Darwin and Lamarck were virtually ostracized for their views by the Establishments, both scientific and non-scientific, of those days.)

Lamarck was wrong in his notion of the evolutionary mechanism, but his book made the concept of evolution well known in the scientific world and it inspired others to find a perhaps more workable mechanism.†

The man who turned the trick was the English naturalist Charles Robert Darwin (grandson of Erasmus Darwin), who spent nearly twenty years gathering data and polishing his argument. This he

† Antievolutionists usually denounce evolution as "merely a theory" and cite various uncertainties in the details, uncertainties that are admitted by biologists. In this, the antievolutionists are being fuzzy-minded. That evolution has taken place is as nearly a *fact* as anything non-trivial can be. The exact details of the mechanism by which evolution proceeds, however, remain theoretical in many respects. The mechanism, however, is not the thing. Thus, very few people really understand the mechanism by which an automobile runs, but those who are uncertain of the mechanism do not argue from that that the automobile itself does not exist.

did, first, because he was a naturally meticulous man. Secondly, he knew the fate that awaited anyone who advanced an evolutionary theory, and he wanted to disarm the enemy by making his arguments cast-iron.

When he published his book On the Origin of Species by Means of Natural Selection in 1859, he carefully refrained from discussing man in it. That didn't help, of course. He was a gentle and virtuous person, as nearly a saint as any cleric in the kingdom, but if he had bitten his mother to death, he couldn't have been denounced more viciously.

Yet the evidence in favor of evolution had kept piling up. In 1847 the largest of the anthropoid apes, the gorilla, was finally brought into the light of European day, and it was the most dramatic ape of all. In size, at least, it seemed most nearly human, or even superhuman.

Then, too, in 1856 the very first fossil remnants of an organism that was clearly more advanced than any of the living anthropoids and as clearly more primitive than any living man was discovered in the Neander valley in Germany. This was "Neanderthal man." Not only was the evidence in favor of evolution steadily rising, but so was the evidence in favor of human evolution.

In 1863 the Scottish geologist Charles Lyell published The Antiquity of Man, which used the evidence of ancient stone tools to argue that mankind was much older than the six thousand years allotted him (and the Universe) in the Bible. He also came out strongly in favor of the Darwinian view of evolution.

And in 1871 Darwin finally carried the argument to man with his book The Descent of Man.

The antievolutionists remain with us, of course, to this day, ardent and firm in their cause. I get more than my share of letters from them, so that I know what their arguments are like.

They concentrate on one point, and on one point only—the descent of man. I have never once received any letter arguing emotionally that the beaver is not related to the rat or that the whale is not descended from a land mammal. I sometimes think they don't even realize that evolution applies to all species. Their only insistence is that man is not, not, NOT descended from or related to apes or monkeys.

Some evolutionists try to counter this by saying that Darwin never said that man is descended from monkeys; that no living primate is an ancestor of man. This, however, is a quibble. The evolutionary view is that man and the apes had some common ancestor that is not alive today but that looked like a primitive ape when it was alive. Going farther back, man's various ancestors had a distinct monkeyish appearance—to the non-zoologist at least.

As an evolutionist, I prefer to face that fact without flinching. I am perfectly prepared to maintain that man *did* descend from monkeys, as the simplest way of stating what I believe to be the fact.

And we've got to stick to monkeys in another way, too. Evolutionists may talk about the "early hominids," about "Homo erectus," the "Australopithecines," and so on. We may use that as evidence of the evolution of man and of the type of organism from which he descended.

This, I suspect, doesn't carry conviction to the antievolutionists or even bother them much. Their view seems to be that when a bunch of infidels who call themselves scientists find a tooth here, a thigh bone there, and a piece of skull yonder and jigsaw them all together into a kind of ape man, that doesn't mean a thing.

From the mail I get and from the literature I've seen, it seems to me that the emotionalism of the antievolutionist boils itself down to man and monkey, and nothing more.

There are two ways in which an antievolutionist, it seems to me, can handle the man-and-monkey issue. He can stand pat on the Bible, declare that it is divinely inspired and that it says man was created out of the dust of the Earth by God, in the image of God, six thousand years ago, and that's it. If that is his position, his views are clearly non-negotiable, and there is no point in trying to negotiate. I will discuss the weather with such a person, but not evolution.

A second way is for the antievolutionist to attempt some rational justification for his stand; a justification that is, that does not rest on authority, but can be tested by observation or experiment and argued logically. For instance, one might argue that the differences between man and all other animals are so fundamental that it is unthinkable that they be bridged and that no animal can conceivably develop into a man by the operation of nothing more than the laws of nature—that supernatural intervention is required.

An example of such an unbridgeable difference is a claim, for instance, that man has a soul and that no animal has one, and that a soul cannot be developed by any evolutionary procedure. —Unfortunately, there is no way of measuring or detecting a soul by those methods known to science. In fact, one cannot even define a soul except by referring to some sort of mystical authority. This falls outside observation or experiment, then.

On a less exalted plane, an antievolutionist might argue that man has a sense of right and wrong; that he has an appreciation of justice; that he is, in short, a moral organism while animals are not and cannot be.

That, I think, leaves room for argument. There are animals that act as though they love their young and that sometimes give their lives for them. There are animals that co-operate and protect each other in danger. Such behavior has survival value and it is exactly the sort of thing that evolutionists would expect to see developed bit by bit, until it reaches the level found in man.

If you were to argue that such apparently "human" behavior in animals is purely mechanical and is done without understanding, then once again we are back to argument by mere assertion. We don't know what goes on inside an animal's mind and, for that matter, it is by no means certain that our own behavior isn't as mechanical as that of animals—only a degree more complicated and versatile.

There was a time when things were easier than they are now, when comparative anatomy was in its beginnings, and when it was possible to suppose that there was some gross physiological difference that set off man from all other animals. In the seventeenth century the French philosopher René Descartes thought the pineal gland was the seat of the soul, for he accepted the then-current notion that this gland was found only in the human being and in no other organism whatever.

Alas, not so. The pineal gland is found in all vertebrates and is most highly developed in a certain primitive reptile called the tuatara. As a matter of fact, there is no portion of the physical body which the human being owns to the exclusion of all other species.

Suppose we get more subtle and consider the biochemistry of organisms. Here the differences are much less marked than in the physical shape of the body and its parts. Indeed, there is so much similarity in the biochemical workings of *all* organisms, not only if we compare men and monkeys, but if we compare men and bacteria, that if it weren't for preconceived notions and species-centered conceit, the fact of evolution would be considered self-evident.

We must get very subtle indeed and begin to study the very fine chemical structure of the all but infinitely versatile protein molecule in order to find something distinctive for each species. Then, by the tiny differences in that chemical structure, one can get a rough measure of how long ago in time two organisms may have branched away from a common ancestor.

By studying protein structure, we find no large gaps; no differences between one species and all others that is so huge as to indicate a common ancestor so long ago that in all the history of Earth there was no time for such divergence to have taken place. If such a large gap existed between one species and all the rest, then that one species would have arisen from a different globule of primordial life than that which gave birth to all the rest. It would still have evolved, still have descended from more primitive species, but it would not be related to any other earthly life form. I repeat, however, that no such gap has been found and none is expected. *All* earthly life is interrelated.

Certainly man is not separated from other forms of life by some large biochemical gap. Biochemically, he falls within the Primates group and is not particularly more separate than the others are. In fact, he seems quite closely related to the chimpanzee. The chimpanzee, by the protein structure test, is closer to man than to the gorilla or orangutan.

So it is from the chimpanzee, specifically, that the antievolutionist must protect us. Surely, if, in Congreve's words, we "look long upon the monkey," meaning the chimpanzee in this case, we must admit it differs from us in nothing vital but the brain. The human brain is four times the size of the chimpanzee brain!

It might seem that even this large difference in size is but a difference in degree, and one that can be easily explained by evolutionary development—especially since fossil hominids had brains intermediate in size between the chimpanzee and modern man.

The antievolutionist, however, might dismiss fossil hominids as unworthy of discussion and go on to maintain that it is not the physical size of the brain that counts, but the quality of the intelligence it mediates. It can be argued that human intelligence so far surpasses chimpanzee intelligence that any thought of a relationship between the two species is out of the question.

For instance, a chimpanzee cannot talk. Efforts to teach young chimpanzees to talk, however patient, skillful, and prolonged, have always failed. And without speech, the chimpanzee remains nothing but an animal; intelligent for an animal, but just an animal. With speech, man climbs to the heights of Plato, Shakespeare, and Einstein.

But might it be that we are confusing communication with speech? Speech is, admittedly, the most effective and delicate form of communication ever conceived. (Our modern devices from books to television sets transmit speech in other forms, but speech still.) —But is speech all?

Human speech depends upon human ability to control rapid and delicate movements of throat, mouth, tongue, and lips, and all this seems to be under the control of a portion of the brain called "Broca's convolution." If Broca's convolution is damaged by a tumor or by a blow, a human being suffers from aphasia and can neither speak nor understand speech. —Yet such a human being retains intelligence and is able to make himself understood by gesture, for instance.

The section of the chimpanzee brain equivalent to Broca's convolution is not large enough or complex enough to make speech in the human sense possible. But what about gesture? Chimpanzees use gestures to communicate in the wild—

Back in June 1966, then, Beatrice and Allen Gardner at the University of Nevada chose a one-and-a-half-year-old female chimpanzee they named Washoe and decided to try to teach her a deaf-and-dumb language. The results amazed them and the world.

Washoe readily learned dozens of signs, using them appropriately to communicate desires and abstractions. She invented new modifications which she also used appropriately. She tried to teach the language to other chimpanzees, and she clearly enjoyed communicating.

Other chimpanzees have been similarly trained. Some have been taught to arrange and rearrange magnetized counters on a wall. In so doing, they showed themselves capable of taking grammar into account and were not fooled when their teachers deliberately created nonsense sentences.

Nor is it a matter of conditioned reflexes. Every line of evidence shows that chimpanzees know what they are doing, in the same sense that human beings know what they are doing when they talk.

To be sure, the chimpanzee language is very simple compared to man's. Man is still enormously the more intelligent. However, Washoe's feat makes even our ability to speak differ from the chimpanzee's in degree only, not in kind.

"Look long upon a monkey." There are no valid arguments, save those resting on mystical authority, that serve to deny the cousinship of the chimpanzee to man or the evolutionary development of Homo sapiens from non-Homo non-sapiens.

10.   O Keen-eyed Peerer into the Future!

I do a lot of after-dinner speaking and also a lot of article writing for the general magazines. In a substantial fraction of the work I get, I am asked to speak on one aspect or another of the future. I have spoken or written, in the not-too-distant past, on the future of such aspects of society as direct-mail advertising, the space effort, amusement parks, supermarkets, sterile disposable devices, and screw machines.

And what makes me such an expert on the future? What are my credentials?

I am a science fiction writer. Nothing more than that.

How respectable science fiction has grown! How almost awe-inspiring a science fiction writer has become merely by virtue of being a science fiction writer. And why? Upon what meat have these, our writers, fed that they are grown so great?

Mostly, it is the predictive aspect of science fiction that has brought about the change. We have been good at predicting.

This is something I have discussed before ("Future? Tense!" in *From Earth to Heaven,* Doubleday, 1966), but that was nine years ago and I have done some thinking on the matter since. What I want to do now is to discuss the matter of prediction in s.f. more systematically (though with some overlap) and present for your consideration the Three Laws of Futurics.*

To begin with, I wish to deny that accurate prediction is the chief concern of the science fiction writer, or even an important minor concern. Nor should it be.

* Everything has Three Laws: Motion, Thermodynamics, Robotics. Why not Futurics as well?

The science fiction writer is a *writer* first and foremost, and his chief and overriding concern, if he is an honest practitioner of his craft, is to turn out a good story and bend everything else to that end. His second concern, since he is also a human being with human needs, is to write the kind of good story that will sell and help him earn an honest living.

If, in the process of writing a good story and earning an honest living, the science fiction writer also manages to make a prediction that eventually seems to come true, so much the better—but it remains a more-or-less accidental by-product of what it is he is doing.

And yet the accurate prediction takes place in science fiction far more often than one would expect from sheer chance. But why not? The science fiction writer, in working out his societies of the future, must base them, consciously or unconsciously, on the society of the present and in doing so automatically evolves a rational way of going about it. In short, whether he knows it or not, he makes use of the Three Laws of Futurics.

Of these, the First Law of Futurics may be expressed as follows: "What is happening will continue to happen." Or, to express it in another fashion, "What has happened in the past will happen in the future." (If it occurs to you that this sounds very much like the old bromide "History repeats itself," you are right. My entire *Foundation* trilogy was consciously guided by the First Law.)

To go into detail on how the First Law works, however, I will take two examples from my own stories: one (mentioned briefly in "Future? Tense!") in which I deliberately violated the First Law, and one (not mentioned in the earlier article) in which I observed it.

First, the violation—

In the spring of 1953 Mount Everest was much in the news. After thirty years of trying, a seventh attempt to scale the mountain had failed.

Yet each successive expedition had learned from its predecessors and each was making use of progressively more sophisticated equipment. By the First Law, we could assume that learning and sophistication would continue and that, therefore, Mount Everest would eventually be climbed.

To attempt to predict the exact day when it would be climbed or the name of the climber or any of the other fine details, would have been, of course, not futurism but fortunetelling; and with fortunetelling the science fiction technique has nothing to do.

In the spring of 1953 I wanted to write a little story about Mount Everest and I could find nothing interesting in the First Law prediction that it would be successfully scaled. If it was, where's the story?

I wanted instead to set up some interesting condition that would cause the prediction *not* to come true. I wanted to find a story in something that was a deliberate violation of the First Law.

(This is not necessarily a bad thing to do. The First Law of Futurics, unlike the First Law of Thermodynamics, *can* be violated. Suppose I were writing a story in 1900 about a future that involves rocket travel. From the fact that man had learned to command steadily greater speeds over the past century, I might assume that First Law predicted that eventually man would achieve a speed of 500,000 kilometers per second. In order to compose an interesting story I would therefore violate the prediction by imagining some sort of cosmic speed limit at 300,000 kilometers per second. It would have been wonderful to do so, for in 1905 Einstein worked out just such a speed limit and made it stick.)

I could invent a number of reasons to abort the inevitability of the scaling of Mount Everest. There could be a sheer, glassy precipice up the last five hundred feet into which picks could not gouge. There might be a mysterious force field blocking off the peak. There might be a layer of poisonous gases six miles up in the air, one that touches ground level only at the peak of the highest mountain.

The aborting effect that I happened to choose was that the Abominable Snowmen really existed and they were actually Martians who had established an observation base on Earth in order to keep an eye on our planet. Naturally, they saw to it that intruding Earthmen in mountaineering costume were either turned back or disposed of.

The story, named "Everest," was only one thousand words long and was sold to a magazine on April 7, 1953, for $30.

You can be sure that I considered the presence of Martians on top of Mount Everest a just-about zero-probability event and was certain that my "prediction" was a false one and that the mountain would be climbed. (Of course, I must admit that in 1900 I would have considered a cosmic speed limit a just-about zero-probability event, too.) Still, I was reasonably certain that the mountain would remain unscaled for a while, anyway, or at least until my story was published.

As it happened, I lost the gamble. At 11:30 A.M. on May 29, 1953, less than two months after my sale, Edmund Hillary and Tenzing Norgay reached the topmost peak of Mount Everest and, needless to say, found neither Martians nor Abominable Snowmen. They had outdated my story before it was published.

Still, the magazine publishers were not about to throw away thirty dollars (and, in those days, I was not about to return the money either) and the story was published anyway. It appeared in the December 1953 issue of *Universe Science Fiction*, which hit the stands in October. I was therefore in the position of having predicted that Mount Everest would never be climbed, five months *after* it had been climbed.

Not one of my more luminous accomplishments!

I had better luck with a much earlier story, "Trends," which had been written a month before my nineteenth birthday and sold a month after my birthday. It appeared in the July 1939 issue of *Astounding Science Fiction*.

It dealt with the first flight around the Moon and back (no landing). I placed the first abortive attempt in 1973 and the second and successful one in 1978. Since the real Moon circuit took place, successfully, in 1968, you can see that I was a decade too conservative.

At the age of not-quite-nineteen I knew nothing about rocket engineering and my notions about what a first flight to the Moon would be like were ludicrously wrong in every respect. There was no government involvement, no military involvement. There were no computers, no mid-course corrections, no preliminary orbital flights, no docking maneuvers, no Russians.

Just to show you how far off I was, I realized dimly that the rocket ship could not be launched in New York City, so I had it launched elsewhere—at the edge of the known world, as I knew it. I had the ship launched across the Hudson River, near Jersey City.

Frankly, it was a terrible story, but no one complained at the time, and it has been anthologized five times (the fifth time in 1973). —You see, all that stuff about rocket ships was not the point. The nub of the story was that there existed great resistance to the matter of space exploration on the part of a large fraction of the population. My rocket-ship inventor was beaten down by this resistance and was driven underground.

This was the first time, in any science fiction story, that resistance to space exploration had been pictured. Until then, science fiction writers either ignored public reaction to space flight or assumed it to be wildly enthusiastic—not only before "Trends," but after "Trends" as well. (To be sure, H. G. Wells had pictured a rocket ship being mobbed in one of his stories, but that was after a futuristic war so that there was then good reason to hate and fear rocket ships. In my story, there was resistance to the very *notion* of space exploration.)

What, then, made this not-quite-nineteen-and-naïve-for-his-age youngster see something that so many older and thicker heads did not see, either before or after? I'll tell you—

I was going to Columbia University at the time and you will not be thunderstruck to hear that I could not afford the tuition. So I scrounged money where I could and for $15 a month I worked for a sociology professor who was doing a book entitled *Social Resistance to Technological Change*.

I had to gather and type the references for him and in doing so, I discovered that there had been embittered resistance to every single significant technological change that had rippled the smooth current of human society—from the discovery of writing to the attempt to build a practical heavier-than-air flying machine.

At once I applied the First Law of Futurics and said to myself, "If that has always happened, it will continue to happen and there will be resistance to space exploration." And I wrote "Trends."

The real question is not why I saw this point, but why the rest of the world did not. The answer is embodied in the Second Law of Futurics which is, "Consider the obvious seriously, for few people will see it."

Surely, I don't have to elaborate on that point to a science fiction audience. It *is* obvious and *was* obvious and *has been* obvious for a long time that increasing population would produce overwhelming problems for mankind, yet most people have steadfastly looked the other way. It *is* obvious and *was* obvious and *has been* obvious for a long time that the oil resources of the world were sharply limited and that the result of allowing that limit to overtake us, unprepared, would be disastrous, yet most people steadfastly looked the other way.

In fact, those, like myself, who kept persistently pointing out the obvious were denounced as "doomcriers" and were shrugged off.

Yet such obvious items are not ignored by science fiction writers. Forced by the professional necessity of considering many possible futures, they apply the Second Law to the construction of dramatic stories—and out of the obvious they make themselves awe-inspiring keen-eyed peerers into the future.

The earliest overpopulation story I personally remember having read and the one which first started me thinking about the inevitability of misery if our population policy continued unchanged was "Earth, the Marauder," by Arthur J. Burks, which ran in the July, August, and September 1930 *Astounding Stories*.

The earliest end-of-fuel story I personally remember having read and which first started me thinking about the inevitability of misery if our fuel policy continued unchanged was "The Man Who Awoke," by Laurence Manning in the March 1933 *Wonder Stories*.

We've been getting these warnings, then, in science fiction, for forty years and more; and yet all our terribly clever statesmen and leaders continue to be caught by "population crises" and "energy crises" and to act as though such crises came out of the woodwork, unheralded, just two days before.

(And yet science fiction magazines were stigmatized as "silly escape literature" for decades. Some "escape"! We science fiction crackpots escaped into a world of overpopulation, oil shortage, and so on—and had the privilege of agonizing for forty years over matters on which all those practical men who read mainstream literature are just beginning to focus their bleary eyes.)

But let's go on. In applying the First Law of Futurics, we mustn't make the mistake of supposing that the "continue to happen" will form an absolutely smooth curve with no landmarks anywhere in it.

No, the curve is bumpy, and sometimes the bumps are sharp and unexpected ones.

There is no way of predicting when a bump will take place, or how sharp it will be, or what it's nature will be—there we are in fortunetelling again. However, it is important to watch carefully for such a bump and to take it into account in foretelling the future.

And, as it happens, science fiction writers, who must devise new societies by the nature of their profession, are often better equipped to see and interpret those bumps than scientists (let alone laymen) are.

In 1880 you might argue by First Law that since mankind was steadily increasing its ability to exploit the energy resources of Earth, it was inevitable that it eventually find some source unknown in 1880—but what it might be, you could not reasonably foretell.

In 1900, when nuclear energy had been discovered as existing, First Law would make it clear that since technology was growing steadily more sophisticated, it would continue to do so and that nuclear energy would be bent to the practical needs of man.

H. G. Wells made this assumption at once and wrote stories of atomic bombs in 1902. Many honest and brilliant scientists—even Nobel laureates in physics—remained convinced right into the 1930s, however, that nuclear energy would never be tamed. It was in this particular spot that science fiction writers proved spectacularly more right than scientists, and it is why scientists today, having been embarrassed by this fact, are now more prone to use the First Law and be generous with their predictions.

Yet if scientists remained conservative on the subject of nuclear energy, it was because through the 1930s there seemed no clear route toward its taming. As late as 1933 Manning could write his story of fuel exhaustion without considering the possibility of nuclear energy as a substitute.

However, in 1939, when uranium fission was announced, one physicist (Leo Szilard) saw the inevitability of a nuclear bomb at once, but so did a lot of science fiction writers (particularly John Campbell) since it was their business to see such things.

The result was that while the United States was working on a nuclear bomb in deep, deep secrecy, science fiction writers all through World War II wrote freely of nuclear bombs and their consequences. (I never did, by the way. I was busy writing my *Foundation* stories and my robot stories and I thought that nuclear bombs were old stuff and not worth wasting time on. Another one of my luminous accomplishments!)

Finally, Cleve Cartmill's "Deadline," which appeared in the March 1944 *Astounding Science Fiction,* roused American security officers. They interviewed John Campbell, the editor of the magazine, on the matter and found that nothing could be done about it. The extrapolation from uranium fission to the nuclear bomb was easy and inevitable (for science fiction writers) and putting any sudden stop to science fiction stories on the subject would give the whole thing away and destroy all secrecy.

To repeat, it was the prediction of the nuclear bomb that most astonished the outside world and that most contributed to the respectability of science fiction—and yet it was so easy a prediction that it deserves no admiration whatever. The outside world should rather have marveled at its own stupidity than at our wisdom.

To achieve particularly important predictions it helps to make use of the Third Law of Futurics, which can most simply be stated as: "Consider the consequences." The prediction of a gadget is easy enough, but what will happen to society when such a gadget becomes established?

To quote a passage from my earlier article "Future? Tense!": ". . . the important prediction is not the automobile, but the parking problem; not radio, but the soap opera; not the income tax, but the expense account; not the Bomb, but the nuclear stalemate."

(My esteemed and long-time friend Frederik Pohl paraphrased this passage from memory in an editorial in *Galaxy* and, unable to remember where he had read it, or who had said it, had prefaced it with "A wise man once said—." When I promptly informed him where he had read it and explained that he had inadvertently called his esteemed and long-time friend Isaac Asimov "a wise man," he was exceedingly chafed, you may be sure.)

The most marvelous example of detecting a consequence that utterly escaped all the great leaders of the world was in "Solution Unsatisfactory," by Anson Macdonald (Robert A. Heinlein), which appeared in the May 1941 *Astounding Science Fiction*. Heinlein predicted the Manhattan Project and the development of a nuclear weapon that put an end to World War II. That was easy. But he also went on to predict the nuclear stalemate, which must have been exceedingly hard to do, since as far as I know, no one else did at the time.

The Third Law can, of course, serve one of the important functions of science fiction—satire. You can consider the consequences and pick a low-probability one that you can make sound so logical as to throw a lurid light on man's folly. The aforementioned Frederik Pohl is very good at that and has written a number of stories designed to show the ridiculous—but logical—consequences that could be brought about by continuing present trends.

I myself have not generally indulged in satire, not being a satiric person by nature. I occasionally manage to do so, however. For instance, I wrote a satiric article for a general-circulation magazine in which I attempted in part to deal with the problem of inflation, with the Third Law in mind. Here are the consequences I pretended to favor—

Consider inflation, for instance— There's a problem that has now become serious. Prices are going up in such a way that misery and suffering aren't confined to poor people who are used to it. Instead, well-to-do people like you and me are beginning to suffer and that is both pitiful and unjust.

Finding a solution did, I admit, take me a little time, because I don't know anything about economicks (if that's how the word is spelled). Fortunately, I recently heard a Cleveland banker discussing some graphs which indicated trends over the next few years. Being a banker, he knew all about ekonomics.

He indicated an upward-sloping line (it meant something; either gross national product or women's hemlines, I'm not sure which) and said that the line was satisfactory but assumed 4 per cent unemployment. "It would be even better," he said, "if we could have 5 per cent unemployment, because that would keep inflation within bounds."

I experienced a blinding flash of illumination. Unemployment was the solution to inflation! The more people unemployed, the fewer people there would be with money. With less money to be thrown around foolishly, there would be just no point in raising prices, so inflation would be solved. I was very glad I had listened to that clever ecconomist.

Now the problem is: How do we go about getting enough unemployed?

The trouble is that it's not a popular occupation and there are practically no volunteers. That's not surprising in view of the contempt with which the profession of unemployment is viewed. How many times have you said to a friend, "Why don't those bums get off welfare and find a job?" (Which is exactly what you don't want them to do, actually, if you are against inflation.)

But view the situation logically. You, with your haughty executive position and your large salary, are contributing to inflation every day, while those poor souls with holes in their shoes, sipping at their wine bottles along skid row, are fighting inflation with desperate intensity. How can you feel contempt for them, then? Which of you deserves more from society?

If we are to defeat inflation, we must recognize the unemployed as our front-line fighters in the battle against that scourge, and we must give them the recognition they should get.

We do so to a limited extent, to be sure. We pay them unemployment benefits or welfare. The pay isn't much; it can't be. If we pay the unemployed a lot, inflation takes over.

But if the pay must be little, must it be accompanied with such open disapproval? Money isn't everything, you know, and any unemployed person would find his pittance quite enough if only it were accompanied by the gratitude he so richly deserves.

What's wrong with greeting these hard-working and long-suffering soldiers of the front-line trenches in the war against inflation with a kindly word, a slap on the back? Let them know we're behind them and think highly of them—but, of course, you must not give them any money. It's essential not to give them money.

The government can help, too. Campaign ribbons for unemployment service can be handed out. The pewter cross with soupspoon clusters can be awarded to those who are unemployed above and beyond the call of duty. The patriotism of certain minority groups who contribute more than their share to the struggle should be recognized. There should be recruiting posters: UNCLE SAM WANTS YOU TO QUIT YOUR JOB!

Men and women would flock to the unemployment colors. The 5 per cent mark would be attained easily—nay, exceeded, for Americans do not shirk their patriotic duty.

And inflation would be brought to a halt!

By way of the Third Law I suppose I am satirizing our economic system; or our callous attitude toward the unemployed; or our romanticizing of war. I'm not sure which, actually, for I only write; I don't analyze.

Whatever the object of the satire, however, it proved too strong. The general magazine which accepted the article asked me to remove this passage and substitute something else. Since it promptly occurred to me I could use the passage otherwise, as I just have, I agreed to do so.

The rejection of the passage is important. One of the difficulties in prediction is that predicting the obvious is sometimes politically and socially dangerous. People don't want their comfort disturbed or their prejudices ridiculed. They don't want to be told that they should sacrifice some of what they have for the poor today or for their own descendants tomorrow. They don't want to be laughed at for their folly. What they want to be told, above all, is that "everything is all right and you don't have to worry."

And by and large that is exactly what they are told, and no one can bring himself to mention this or that potential discomfort till it has become too enormous and overpowering to be denied any longer.

But my passage on inflation *can* be published in a science fiction magazine, and so can anything else (provided it is written well enough) no matter how discomforting it may be to the comfortable or how unpalatable to the social gourmet.

It is the very nature of science fiction to consider the discomforting if that is where the task of extrapolating social and scientific trends takes us; and the wonderful thing about the science fiction reader is that he will accept the discomfort and look it in the face.

If we could get the whole world to do that, there might be a chance for humanity yet.

D—MATTER

11. The Mispronounced Metal

As a die-hard one-worlder, I scorn the way people quarrel over languages. What difference does it make whether you speak one language or another, as long as we all learn some *one* language which we might call "Earth-standard"? Then, if you're not trying to make yourself understood generally, use whatever private code you wish, for goodness' sake.

Naturally, I think Earth-standard should be very close to English. It's not because I happen to know English, but because English is already more widely and commonly spoken than any other language on Earth and is still on the way up.

Consider, then, how lucky I am—I already speak Earth-standard. No wonder I can afford to be lofty about the petty prejudices of those people who don't, and consider their quarrels over languages that are not Earth-standard to be childish.

So you think that I would not be affected by minor dialectical differences in English. If I scorn quarrels over entire languages, I certainly won't be perturbed by a small matter of pronunciation.

Oh, do you? You understand nothing about human nature, then.

Last night I was watching an episode of "The Avengers," which I watch every chance I get since there are few episodes I have seen oftener than a dozen times. —And in this episode, one of the characters casually referred to a "school schedule," pronouncing it "skool shedule."

I was rocketing out of my chair at once, crying out incoherently something that would have been like this if I could have maintained my cool. "Shedule?" I was trying to say. "*She*dule? Why not say 'shool shedule'? Why not say 'sholar' for 'scholar,' and 'sheme' for 'scheme,' and 'shizophrenia' for 'schizophrenia,' and 'Shenectady' for 'Schenectady'? Only in German is 'sch' pronounced 'sh' as in 'schnitzel' and 'Schubert.' You hear me? You hear me?"

They didn't hear me. I missed a full five minutes of the program and it did me no good. Worse yet, I do this everytime I hear anyone mispronounce "schedule" in that jackass way, and it never does me any good.

It happens in science, too. What do you call that nice, shiny white metal they use to make sidings and airplanes out of? Aluminum, right? Aluminum, pronounced "uh-LOO-mih-num," right? Anybody knows that!

But do you know how the British spell it? "Aluminium," pronounced "Al-yoo-MIH-nee-um." Ever hear anything so ridiculous? The French and Germans spell it "aluminium," too, but they're foreigners who don't speak Earth-standard. You'd think the British, however, using *our* language, would be more careful.

Oh, well, though there's nothing I can do about "shedule," there's something I can do about "aluminium." I can write an article about the metal and spell it ALUMINUM all the way through.

It started in ancient times in connection with dyes. There were very few decent dyes in ancient times. Some substances had the color one would expect of a dye, but were useless because they wouldn't stick to the fabric. They would just give it a faint tinge and then wash out the next time the garment was beaten with sticks at the river bank.

At least as far back as 2000 B.C. in Egypt, however, it was discovered that if the garment were first boiled with solutions of certain colorless substances and then washed and *then* boiled with the dye—the dye would stick. Apparently, the colorless substance sticks to the fabric and the dye sticks to the colorless substance.

Such an intermediate compound is called a "mordant," from a Latin word meaning "biting," because the mordants used were bitter solutions that hurt if they found the exposed nerves in small cuts and abrasions. The particular mordant most commonly used was called "alumen" by the Romans, a word which seems to be related to Greek words meaning "bitter." We call it "alum."

Chemically, an alum is any of a large class of double-salts in which a sulfate group is attached to two different metals. The variety that is most common (and perhaps most frequently used by the ancients) is common alum, or potassium alum. It is actually potassium aluminum sulfate and if you want its chemical formula it is $K_2SO_4 \cdot Al_2(SO_4)_3 \cdot 24H_2O$.

We meet up with alum in the styptic pencil used for minor cuts in shaving. The word "styptic" is from a Greek word meaning "to contract," because alum causes small blood vessels to contract. After the biting pain of the first touch of the styptic pencil, the bleeding stops. Alum is also called an "astringent," from a Latin word meaning "to draw tight" or "contract."

In the eighteenth century mineralogical chemistry had blossomed and there was an enormous push toward determining the basic constituents of the various rocky substances used by mankind. These basic constituents were termed "earths," largely because they shared the properties that the rocky crust of the Earth had—they did not dissolve in water, nor melt in fire, nor burn in air.

The first to obtain what seemed a simple earth from alum was a German chemist, Johann Heinrich Pott, in 1746. Another German chemist, Andreas Sigismund Marggraf, also reported it in 1754 and went further—he discovered he could obtain the same earth, whatever it was, from various clays. Furthermore, he showed it was a distinct earth, with properties different from the earth obtained from chalk and limestone.

It was customary in those days to give earths that did not already have some common name the ending "a," attached to the stem of the name of the mineral from which it was obtained. The earth that came from alumen was, therefore, named "alumina."

By the end of the eighteenth century the French chemist Antoine Laurent Lavoisier had established modern chemistry and had shown the key role played by oxygen in combustion and in rusting. He maintained that the various earths were made up of some metal in combination with oxygen. The combination was so tight that there did not exist any laboratory methods to break it, so the metal remained unknown.

We could argue through hindsight, of course, that since atoms were held together by electrical forces, the grip might be broken by the use of electrical forces. The chemists of 1800 didn't know about atoms and electrical forces, but methods for producing an electric current were just being devised and chemists were anxious to make use of this new and glamorous phenomenon.

An electric current won't go through the typical mineral, but it will, sometimes, go through the mineral when it is liquefied. In 1807 and 1808 the English chemist Humphry Davy melted certain minerals and passed an electric current through them, obtaining the metals they contained in pure form. In this way, he produced metals such as sodium, potassium, magnesium, calcium, strontium, and barium.

These metals hold on to other atoms so tightly that anything short of an electric current won't pry them loose. Once loose, they have a strong tendency to combine with anything in reach. Naturally, they combine with oxygen from the air. They even snatch oxygen from the water molecule, which is made up of oxygen combined with hydrogen. The hydrogen which is left behind bubbles off and generally catches fire. For that reason, Davy's metals, when kept for use in the chemistry laboratory, are left immersed in a non-oxygen-containing liquid such as kerosene.

Notice, by the way, that Davy's metals all have the same ending. In the 1780s Lavoisier had established the systematic chemical terminology we still use today, and there was international agreement to adopt the Latin ending "um" for metals that were newly discovered and did not already have a common name. The Romans, you see, had used that ending. To them, gold, silver, copper, iron, tin, and lead had been aurum, argentum, cuprum, ferrum, stannum, and plumbum.

The "um" was added to the stem of the name of the mineral from which the metal was obtained. The metal from the mineral baryta (from a Greek word meaning "heavy") was named "bari-um" (the "y" and "i" being equivalent). The metal obtained from the mineral strontianite (named for Strontian, Scotland, where it was found) was named "stronti-um." The metal obtained from the mineral magnesia (named for an ancient Greek town) was named "magnesi-um," and so on.

Through pure chance, many metals retained the "i" from the name of the mineral so that the ending was "ium." However, it was "um" that was the essential ending. Thus, certain metals, discovered in 1748, 1781, and 1802, were named "platinum," "molybdenum," and "tantalum," respectively, names that are kept to this day and which are spelled and pronounced identically in Great Britain and in the United States.

But let us get back to alumina. Could it be broken up by an electric current and the metal obtained? Unfortunately, it couldn't, because neither alumina nor any related compound could be melted at any reasonable temperature and an electric current could not be forced through it.

For a while, Davy had thought he had succeeded, and he named the metal "aluminum" in perfectly correct fashion—the "um" ending placed on the stem of the name of the ore.

Alas, it did not stick. The weight of precedent had moved heavily in favor of the "ium" ending. Since 1802 only one out of the nearly sixty metals that have been discovered received a straight "um" ending and that was "lanthanum." There was therefore a strong push in favor of "aluminium" rather than "aluminum" and this was *wrong*.

No, not because I feel it violates Latin or anything as prissy as that. Consider, though, that until 1880, not a single element had been given an English name of more than four syllables. Why should "aluminium," with five syllables, be introduced? (In fact, try to say "aluminium" three times rapidly and anyone listening to you will burst into laughter.)

Since 1880 multisyllabic elements have been with us, for a variety of reasons. There are seven elements with five syllables now: gadolinium, neodymium, protactinium, americium, californium, mendelevium, and rutherfordium. There is even one element with six syllables: praseodymium. All are uncommon elements, however, that would never choke the mouth of anyone but a professional chemist. But why give five syllables to an element as commonly in everyone's mouth as "aluminum"?

Right? —Right!

Now that that's settled, let's get on to the isolation of the metal in alumina. Davy had failed with an electric current, but what about more conventional methods?

Alumina is made up of aluminum and oxygen (Al_2O_3) held together very tightly. If one could use some element that held on to oxygen even more tightly than aluminum did, it would replace the aluminum, which would then be left behind in its metallic form.

Davy's metals formed even tighter bonds with oxygen than aluminum did, so what about them? Of course, they were dangerous to use, and potassium, the most active, was also the most dangerous. What's more, they were expensive, but, at the time, potassium was least expensive.

The first to try this was a Danish chemist, Hans Christian Oersted. The details don't matter, but in essence, his method was to free the aluminum atoms by replacing them with potassium, using metallic potassium for the purpose.

What Oersted obtained in this way, in 1825, was at best a very impure sample of aluminum, but some was there and he was therefore the first man in the history of the world to set eyes on that silvery metal. In 1827 the German chemist Friedrich Wöhler used a modification of Oersted's method to obtain a somewhat purer sample of aluminum, enough of it in pure enough form to get an idea of its properties.

Those properties turned out to be quite remarkable. For one thing, aluminum was very light for a metal. Whereas a cubic centimeter of iron weighed 7.86 grams, a cubic centimeter of aluminum weighed 2.70 grams, only one third as much.

There were metals that were less dense than aluminum, to be sure. The density of potassium and sodium are 0.86 and 0.97 grams per cubic centimeter, only one third that of aluminum.

But there is a difference. Sodium and potassium are so eager to combine with almost anything that they don't *stay* sodium and potassium very long. And even if they did, they are as soft as wax and can't be used for the usual structural functions of, say, iron.

Aluminum, on the other hand, although it has nearly as great a tendency to combine with other atoms as sodium and potassium have, does not do so in practice. Why not? Well, as soon as it is formed, the aluminum atoms on the surface bind themselves strongly to oxygen atoms from the air. The aluminum oxide so formed remains on the surface, one molecule thick, and forms so tight and coherent a layer that the aluminum atoms underneath aren't touched by additional oxygen even over prolonged periods. The aluminum oxide layer is so thin as to be transparent and aluminum continues to look perfectly metallic and uncorroded.

Indeed, aluminum is far better in this respect than iron is. Iron is less active than aluminum and its atoms have a lesser tendency to combine with oxygen. Iron atoms do combine, though, especially in the presence of water, and when they do so, the iron oxide that forms is, in the first place, orange and shows up as an unsightly rust over the metal. Then, too, the oxide is crumbly and falls away, uncovering more iron atoms that combine with oxygen in their turn.

But in that case, why isn't metallic aluminum found in nature if, once formed, it remains metallic? Ah, the catch is in the phrase "once formed." The geological processes that formed the crust of the Earth scattered aluminum in the form of widely dispersed atoms, all of which combined with oxygen and other atoms. It is only man who has concentrated aluminum atoms in bulk so that those on the surface could protect those beneath.

Of course, iron is stronger than aluminum if we consider cylinders of given size. If we consider weight, however, then an aluminum cylinder of a given length would be greater in diameter than an iron cylinder of the same weight and length. The greater strength of iron would be not nearly as pronounced, then, weight for weight.

Next, consider the way in which metals conduct electricity. The best conductors are silver, copper and gold, in that order. The resistivity for these three, in microhm-centimeters at 20°C, is 1.59, 1.72, and 2.44, respectively. Since copper is the most available of the three, and is better than gold and not too much worse than silver, it is the preferred material for electrical wiring. Anything else would either increase the expense, increase the loss of electrical energy as heat, or both.

Well, not quite anything else. Consider aluminum, which has a resistivity of 2.82. Aluminum is only $\frac{3}{10}$ as dense as copper. If the same weight of aluminum and of copper were used to form wires of given length, the aluminum wire would have a cross-sectional area $3\frac{1}{3}$ times that of copper and the aluminum wire would then have only half the resistivity of the copper wire.

In short, weight for weight, aluminum is the best electrical conductor. And the same goes for the closely allied property of heat conductivity.

Aluminum also has the very unusual property of retaining its metallic and silvery shine when reduced to a fine powder. Aluminum gleams brightly while powdered metals of other kinds tend to be black. If you suspend the aluminum powder in some appropriate dispersing medium, you have aluminum paint.

And, of course, as in the case of many other metals, aluminum can be beaten into thin layers, so you can have aluminum foil lighter and shinier than most metal foils.

Think of those uses, then. Think of aluminum's lightness, its strength, its non-corrodability, its electrical conductivity, and so on and so on. Surely, the possibilities are delightful—unless the metal should happen to be rare. The best metal in the world is of no use for most purposes if it is so rare that it can be obtained only in small quantities and then only at great expense.

Well, relax, aluminum is not a rare metal at all. For every gram of copper in the Earth's crust, there are 1,100 grams of iron and 1,800 grams of aluminum. The discrepancy is even greater in terms of atom numbers. For every atom of copper in the Earth's crust, there are 1,250 atoms of iron and 4,750 atoms of aluminum.

Aluminum is, actually, the most common metal in the Earth's crust. There are nearly four times as many aluminum atoms all about us as iron atoms—the next most common metal.

And yet all is not well. The trouble lies in the difficulty of getting those aluminum atoms to let go of oxygen atoms. Iron oxides can be heated with plentiful, cheap, and safe carbon atoms in the form of coke or charcoal, and metallic iron is with us at once. For aluminum oxides, carbon atoms aren't enough as oxygen-grabbers. In the 1820s it was the exceedingly dangerous and expensive potassium that had to be used, and even then the aluminum obtained was impure.

The first pure aluminum was prepared in 1854 by the French chemist Henri Sainte-Claire Deville. Sainte-Claire Deville had worked out methods for producing metallic sodium in larger quantities than had hitherto been possible. This meant that metallic sodium became considerably cheaper than potassium. It was not quite as active as potassium and was therefore safer to use, and it was still active enough to replace the aluminum as a gripper of oxygen.

Sainte-Claire Deville repeated Wöhler's method for preparing aluminum, with the substitution of sodium for potassium. Using generous quantities of this now-readily-available material, he produced a quantity of pure aluminum.

But just because sodium was cheaper than it had been didn't mean it was *cheap*. By Sainte-Claire Deville's method, pure aluminum remained an expensive and, indeed, a semiprecious metal. It cost $10 a pound through the 1870s, and $10 in the 1870s meant many times what it does now a century later.

Napoleon III, Emperor of the French, searching for a properly imperial gift for his infant son, gave him an aluminum rattle. And the Americans, in 1884, completed the Washington Monument by placing nothing less lavish than an aluminum tip on it. None of your plebeian gold.

The delightful properties of aluminum were known, of course, but it was also quite certain that it would remain an expensive metal so long as sodium or potassium were needed for its preparation. Ah, if only electrical methods could be used in preparing aluminum directly, rather than in preparing sodium as the middleman—

One person interested in aluminum production was the American chemist Frank Fanning Jewett, who had studied in Germany under Wöhler. In 1885 he was teaching chemistry to the senior class at Oberlin College. In discussing the properties of aluminum in class, he sighed and said that anyone who could devise a practical method for preparing aluminum cheaply would surely make a fortune.

In the class was young Charles Martin Hall. Fired up, he decided to devote himself to the task of finding such a cheap method. He set up a chemical laboratory in a woodshed, put together some electrical batteries, and got to work.

He needed an aluminum compound in liquid form. Alumina was no good, for it melted at a temperature of 2,050°C and neither Hall nor anyone else could work cheaply at that temperature.

Of course, some aluminum compounds can be easily melted. One such is sodium aluminum fluoride, which is found in nature as the mineral cryolite.

The "cryo-" prefix is from a Greek word meaning "icy cold" and it is a fitting name for a number of reasons. First, it has the appearance of ice and an index of refraction almost exactly like that of water, so that it seems to disappear when placed in water as ice does (though, of course, this is only seeming, since the cryolite neither melts nor dissolves in the water). It does melt at quite a low temperature, however. The heat of a candle will do the job, so that it almost seems to be a high-melting ice. Finally, the only good natural source of cryolite, discovered in 1794, is near Ivigtut, on the west coast of Greenland's southern tip, which is another association with ice.

Could an electric current passed through molten cryolite liberate the aluminum atoms present in its structure? To a limited extent, but not very well.

Hall made the crucial discovery, however, that aluminum oxide would dissolve in molten cryolite. The dissolved aluminum oxide would then be, in effect, in liquid form and at a temperature that Hall could easily handle in his woodshed. He pushed in the electric current—and out came the aluminum.

On February 23, 1886, Hall rushed into Jewett's office and in his hand were small nuggets of pure aluminum. (Those nuggets are still preserved by the Aluminum Company of America in Pittsburgh as the aluminum "crown jewels.") At the time, Hall was only a few months past his twenty-second birthday.

The process was soon put into production. Hall had his legal problems, but they were all straightened out and he ended up, as Jewett had foreseen, making a fortune. And the price of aluminum plummeted. By 1900 it was no longer either rare or expensive.

And here's something that's odd. While Hall was working out his electrolytic method of preparing aluminum, another chemist in France, Paul Louis Toussaint Héroult, was working out precisely the same method, molten cryolite and all.

Both Hall and Héroult have names beginning with H, but what is more remarkable is that both men were born in 1863, so that both were in their early twenties when they made their discoveries. And it further happens that both died in 1914, each one month after his fifty-first birthday. A peculiar coincidence.

Of course, you may think that the Hall-Héroult process had a serious deficency in requiring cryolite. If the only source of that mineral is in southwestern Greenland (and no other important source has ever been discovered), then that limits the availability of aluminum. And when the cryolite is used up, gone is the aluminum.

Actually, the cryolite doesn't get used up rapidly; a little goes quite a long way. Then again, the aluminum industry no longer uses natural cryolite from Greenland; it makes it out of more common substances and the supply of this synthetic cryolite will last indefinitely.

(Of course, electricity is still a considerable item. Since aluminum is prepared by an electric current, and iron by heating with coal, aluminum remains more expensive than iron.)

Almost immediately after aluminum became cheap, it showed what it could do in a startlingly new fashion. Here's the way it came about—

Mankind learned to fly in 1783 with the construction of the first balloons capable of lifting human beings into the air. For over a century, balloons kept growing more elaborate, but they were essentially powerless, drifting mechanisms going wherever the wind carried them.

The method of correcting this was plain. Balloons could be made large enough to lift steam engines or internal-combustion engines along with a crew, and these engines could be hooked up to propellers. The balloon could then be driven against the wind, if desired, just as a steamship can be driven against the ocean current.

To make such a "dirigible balloon" (one, that is, that could be "directed"), however, the balloon itself should be formed in some streamlined shape; otherwise, too much energy would be expended just overcoming air resistance.

The spherical shape, which was natural for a balloon, was horribly inefficient. What was wanted was a cigar shape, the long axis parallel to the ground, but if a balloon was manufactured with walls of varying strengths in order to make it expand into a cigar shape, it would be both expensive and unsafe.

An alternative was to place the balloon (or balloons) into a cigar-shaped container made of something strong enough to maintain the shape through the normal buffeting of wind and weather and yet light enough to be lifted without costing all the efficiency gained through streamlining.

The German army officer Count Ferdinand von Zeppelin thought that aluminum might fill the bill. He constructed a hollow cigar-shaped aluminum structure, 420 feet long and 38 feet thick, and placed hydrogen-filled balloons within it. Underneath the cigar were two gondolas, each of which contained an engine geared to two propellers. It was the first of a class of vehicles variously called "zeppelins," "dirigibles," and "airships."

On July 2, 1900, Von Zeppelin flew the first airship at speeds of up to twenty miles per hour. It was the *first* powered flight in history* and aluminum made it possible.

And here I might mention one of the less frequently referred to predictions of science fiction. In 1865 Jules Verne published his *From the Earth to the Moon*. At the time, twenty-one years before the Hall-Héroult process, aluminum was still a precious metal, but Verne, thoroughly appreciating the fact that no transportation device intended to be lifted off the ground could be built of anything heavy, had his spaceship built of aluminum!

* The Wright brothers at Kitty Hawk, three and a half years later, demonstrated the first powered flight of a *heavier-than-air* machine.

12. The Uneternal Atoms

I have recently returned from the 31st World Science Fiction Convention, held in Toronto. It was a thoroughly satisfactory experience. The hotel (the Royal York) was splendid; the program was efficiently handled; the attendance was nearly 2,500; and, best of all, at the Sunday night banquet, it was announced that my novel *The Gods Themselves* had won the Hugo.

By the end of that particular day you can well imagine that I was feeling very happy as I stepped into the elevator to go up to my room.

In the elevator were present four people. Two were strangers who were dressed in suits, ties, and tight collars, with glistening new-reaped chins, short hair, and solemn expressions of intense respectability. The other two were fifteen-year-old science fiction fans who had the young and innocent fuzziness of the species.

Unfortunately, although I am somewhat over fifteen years of age, I did not look very respectable myself. I was cold sober, of course, despite provocation, but at the end of the day I tend to have a kind of rumpled look.* Furthermore, my leonine shock of long and graying hair was in more than usual disorder.

As soon as I entered the elevator, one of the strangers said, with what seemed to me to be a glint of disdain in his eye, "What's your club?"

At once the fifteen-year-olds cringed. I knew what was passing through their minds for I had been there once. They were going to be made fun of and humiliated by respectable people who thought science fiction was half-wit nonsense.

With an inner sigh, I assumed my Establishment intonation (which I can manage, with a modicum of effort, though I much pre-

* At the beginning of the day, too.

fer to speak my native Brooklynese) and said, "To what, sir, are you referring?"

"What's that?" he said, pointing to my Torcon II button.

"That," I said, "means that I am attending the 31st World Science Fiction Convention, which is the second of its kind to be held in Toronto; hence Tor-Con-Two."

"Science fiction?" There was a small, tight smile on his face. "What can you possibly do at science fiction conventions?"

I said, "We listen to speeches and panel discussions; we discuss the state of the art among ourselves; we hold a costume party; we introduce notables; we attend a banquet; we hand out coveted awards. In short, sir, science fiction conventions are just like other conventions, except, of course, that those who attend science fiction conventions are *much* more intelligent than those who don't."

And with that, leaving two satisfied fifteen-year-olds behind me, I bounded happily out of the elevator.

But I wasn't just trying to put down a pair of saps, you know; I meant what I said. I write articles for a wide variety of magazines, but these essays right here, the ones that appear originally in *F & SF*, are the only ones in which I never feel the need to pull any punches. —So let's go into the matter of radioactive breakdown.

Most of the common atoms about us are stable. By that I mean that, left to themselves, unimpinged upon by the outer universe, they would retain their structure unchanged, as far as we know, through all eternity. Some atoms, however, even if left to themselves, break down, give off radiation in the form of photons and/or massive particles, and become atoms of another structure. Such atoms are unstable and, because of the radiations they give off, are said to be "radioactive."

Suppose you have a single atom of a particular variety (or a "nuclide") that is radioactive. That it will eventually break down you can be sure, but exactly *when* will it break down? There is no way of telling that. It may break down after five minutes, or after five years, or after five billion years.

But if we can't speak of certainties, we can at least speak of probabilities. Some radioactive nuclides are very unstable, and it is much more likely that a given atom of that type will endure less than five minutes before breaking down than that it will endure more than five minutes. On the other hand, some radioactive nuclides are only slightly unstable, and it is much more likely that a given atom of that type will endure more than five billion years than less.

We can't tell what the probabilities are, *in general*, for a particular nuclide by just observing *one* atom and noting when it breaks down; but we can do it by observing a large number of atoms of a particular nuclide and noting how many break down after one minute, how many after another minute, how many after another minute and so on.

If we do this, we can work out the mean-life—that is, the average life expectancy.

This is not a very unusual concept. It is exactly what is done in the life insurance business. If we consider a particular human being concerning whom we know nothing, we cannot tell whether he will die in five minutes or in fifty years. However, if we study a great many human beings we will find that we can work out, with considerable precision, the fraction that will die within one year, even though we can't possibly tell which particular individuals will make up that fraction. From a study of many human beings we can calculate our life expectancy.

The situation with human beings is very complex, however. The life expectancy can vary with geography, sex, age, social status, past history and so on. The life expectancy of American males is considerably higher than that of Nigerian males and rather lower than that of Swedish males or of American females. The life expectancy of white American males is somewhat higher than that of black American males; that of rural Americans somewhat higher than that of urban Americans; of non-smokers than of smokers; and so on. Again, the life expectancy of fifteen-year-old Americans is considerably greater than that of seventy-five-year-old Americans.

All these complications are not present in the case of atoms. In considering atoms of a particular nuclide, it matters not from what part of the Earth it derives (or from what part of the Universe, as far as we know) or what its surroundings are or how long it has already existed unchanged. The life expectancy is always and forever the same, for atoms neither age nor sicken.

This unvarying life expectancy makes the mathematics of the breakdown process rather simple and it is easy to demonstrate some interesting properties of radioactive atoms from the equations. For instance, it can be shown that for any collection of atoms of a particular radioactive nuclide, any given fraction will always break down in a given time.

Suppose we take the fraction $^{17}\!/_{273}$. Given one pound of radioactive nuclide Q, we will find that $^{17}\!/_{273}$ of that pound will break down in, say, one year. In that case, if we start with two pounds, then we can be sure that $^{17}\!/_{273}$ of two pounds will break down in a year. If we start with a ton, then $^{17}\!/_{273}$ of that ton will break down in a year. In fact, $^{17}\!/_{273}$ of all the Q that exists in the entire Universe will break down in that year.

But since this is true of any fraction, why choose $^{17}\!/_{273}$? Why not choose the simplest one of all, $\frac{1}{2}$? Back in 1904, in fact, the British physicist Ernest Rutherford suggested this be done. Once the life expectancy of a particular radioactive nuclide is worked out from some appropriate set of observations, then from that the time in which one half the atoms of that nuclide will break down (its "half-life") can be calculated.

The half-life is absolutely characteristic of any particular radioactive nuclide. The more nearly stable it is, the longer the half-life; the more unstable it is, the shorter the half-life. And remember, too, that the half-life is not affected by the previous history of the atoms.

Suppose, for instance, that we start with a quantity of atoms of a nuclide with a half-life of one year. In one year, half are gone and only half are left. The half that remain, however, still have a half-life of one year. In another year, then, half of the remaining half are gone and $\frac{1}{4}$ of the original number of atoms are left unchanged. In another year after that, half of the remaining quarter, or $\frac{1}{8}$, are left, and in another year $\frac{1}{16}$, and so on.

If we continue this onward we can see, from the strictly mathematical viewpoint, that the series of fractions gets smaller and smaller in value but never decreases to zero. We can therefore argue (and I have seen it so argued) that although individual radioactive atoms are mortal, the nuclide itself is immortal—that there may be fewer and fewer such atoms with time, but that, however short the half-life of the nuclide, the number never declines to zero.

This is true if we deal with mathematical symbols only or if we start with an infinite number of atoms—not otherwise.

The mathematical equations that describe the course of the break-down of radioactive atoms depend upon a statistical analysis which, in turn, depends upon the presence of so vast a number of atoms that random variations in the behavior of individual atoms cancel out. The smaller the number of atoms being considered, the greater the influence of random variation and the less applicable the equations.

Put it another way, no matter how large a finite number of atoms you begin with, and no matter how long the finite half-life, then, given time enough, you will eventually be down to a single atom, and eventually that last atom will also go. So, provided we neglect that fact that atoms of a particular nuclide may be formed in the course of time, we find that radioactive atoms are uneternal, whether they are considered as individuals or as nuclides.

Since the last few atoms of any nuclide or of any limited sample of a nuclide behave in an increasingly random manner and since no one can tell when the last will go, there is no point in talking about the "life" of a particular nuclide and no one ever does. It is always "half-life."

There may be some point in choosing a fraction considerably larger than $\frac{1}{2}$. Suppose we try $\frac{127}{128}$. If a radioactive nuclide exists on Earth in geologically significant quantities, then even after $\frac{127}{128}$ (99.22 per cent) has gone, the remaining $\frac{1}{128}$ (0.78 percent) will still include a respectable number of atoms to which the equations will apply with good accuracy.

Still, compared with what we began, $\frac{1}{128}$ can fairly be considered a small quantity. It may well be that the kind of techniques evolved to deal with the original quantity would not work with sufficient accuracy when only $\frac{1}{128}$ is left and the feeling would be that the nuclide had declined to trace quantities.

Let's therefore call the time after which only $\frac{1}{128}$ of the original quantity of a particular radioactive nuclide is left its "trace-life."† But why $\frac{1}{128}$? Because $\frac{1}{128}$ is $\frac{1}{2} \times \frac{1}{2} \times \frac{1}{2} \times \frac{1}{2} \times \frac{1}{2} \times \frac{1}{2} \times \frac{1}{2}$, so that the trace-life is exactly seven times as long as the half-life.

Now let's start from another track. The Earth is supposed to have existed as a more or less solid globe of its present size for 4,600-

† I must confess that this term and, indeed, this concept, is original with me. Please hold organized science guiltless of it.

000,000 years. If we call a billion years an "eon" (as is becoming increasingly common), we can reduce the number of zeros with which this article will otherwise be riddled and say, for instance, that the age of the Earth is 4.6 eons.

If, then, any radioactive nuclide has a trace-life of more than 4.6 eons, it will be existing on Earth, even today, in more than traces, even if we must depend only on those atoms originally present on Earth and discount the possibility of new creations since Earth's formation. Any nuclide with a trace-life of more than 4.6 eons would have a half-life of more than 0.66 eons. We can refer to any nuclide with a half-life of more than 0.66 eons, then, as a "long-lived nuclide" with atoms that have come down to us from Earth's beginning.

(To be sure, radioactive nuclides with half-lives of less than 0.66 eons—and even far, far less—exist on Earth, because they are continually being formed. It is not with these Johnny-come-latelies that this article is concerned, however.)

Of the atoms that have existed on Earth from the beginning, most are members of the various stable nuclides that, as far as we know, don't break down at all. There are 264 nuclides of this sort divided among 81 different chemical elements. Again, as far as we know, these are the only 264 nuclides that can possibly be stable, given the laws of physics as they are.

In addition, though, there are atoms that have existed on Earth from the beginning which are members of twenty-one long-lived radioactive nuclides with half-lives that are greater than 0.66 eons. These are listed in Table 8 in order of decreasing half-life.

Some of the half-lives of these long-lived nuclides are inordinately long. The half-life of lead-204 is fourteen billion eons. Even during the 4.6-eon lifetime of Earth, only a tiny fraction of lead-204 has had a chance to break down—only one ten-billionth, in fact. Since there are about six trillion (6,000,000,000,000) tons of lead-204 in the Earth's crust, we can say that during the entire course of Earth's long history, only six hundred tons of lead-204 has broken down.

Of course, if we want to deal in single atoms, we can express this a little more dramatically. So enormous are the number of atoms in any reasonable quantity of material that even if it takes fourteen billion eons for half of them to break down, an appreciable number

TABLE 8 — THE LONG-LIVED NUCLIDES

Nuclide	Half-life (in eons)
Lead-204	14,000,000,000
Calcium-48	20,000,000
Cerium-142	5,000,000
Neodymium-144	5,000,000
Hafnium-174	4,300,000
Platinum-192	1,000,000
Vanadium-50	600,000
Indium-115	600,000
Samarium-149	400,000
Gadolinium-152	110,000
Samarium-148	12,000
Platinum-190	700
Lanthanum-138	110
Samarium-147	106
Rhenium-187	70
Rubidium-87	47
Lutetium-176	21
Thorium-232	13.9
Uranium-238	4.51
Potassium-40	1.30
Uranium-235	0.713

manage to break down in the first second.

Suppose we imagine ourselves to possess a pound of pure lead-204. (This is a piece of good imagination, for ordinary lead is made up of a mixture of lead-204, lead-206, lead-207, and lead-208, with lead-204 the component present in the smallest quantity— 1.48 per cent of the whole. If your fairy godmother gave you a pound of pure lead-204, you could sell it for a fantastically large sum.)

In that pound, if you had it, about 30,000 atoms would be breaking down each second. This rate would continue onward for second after second at an apparently constant rate. Measurements throughout the entire space of man's civilized existence on Earth would not be able to detect any decline in the rate of breakdown. But, of course, the rate would be very slowly declining just the same, and after fourteen billion eons, when half the atoms in the pound of lead-204 had broken down, the rate of breakdown would have declined to 15,000 atoms per second.

And, while we are thinking of lead-204, who is to say that fourteen billion eons is the top value for half-life? It's just that the longer the half-life, the fewer breakdowns per second there are, the feebler the radioactive intensity is, and the harder it is to detect it. If some nuclide were breaking down more slowly than lead-204, it would have a still longer half-life and would be still more feebly radioactive. We might not be able to detect so feeble a radioactivity but it would be there.

Perhaps as we learn to detect feebler and feebler levels of radioactive intensity, we could find half-lives that were longer and longer and end by deciding that *every* nuclide (except perhaps hydrogen-1, the simplest, from which all the others arose since the beginning of the Universe) is radioactive to some more or less infinitesimal degree and that only hydrogen-1 is truly stable.

But let's be reasonable. However numerous 30,000 atoms per second seems to us, out of a pound of lead-204, with its trillion-trillion atoms, it is virtually nothing. We might as well consider lead-204 to be effectively stable, even if it is not ideally so. In fact, in order to avoid trying to make an absolute distinction between radioactive and non-radioactive nuclides—a distinction which may not exist and which may only be illusorily imposed on us by the state of the art—let's say that any nuclide that has lost less than 1 per cent of its mass through radioactive breakdown in the course of the Earth's existence is effectively stable.

For more than 1 per cent of a nuclide to have broken down over the course of the Earth's lifetime, the half-life must be less than 320 eons.

So we have worked ourselves down to a consideration of those nuclides which, through radioactive breakdown in the course of Earth's history, have lost more than 1 per cent of their mass and less than 99.22 per cent of their mass. These have a level of intensity of radioactivity high enough to be considered reasonably substantial and yet not so high as to forestall a useful amount from existing in Earth's crust today.

The number of nuclides that meet the exacting requirements of having a half-life between the limits of 320 eons and 0.66 eons are exactly nine in number, and it is highly doubtful that a tenth member of this exclusive club will ever be discovered. The long-lived radioactive nine are listed in Table 9 in order of decreasing half-life, and the fraction of the original quantity of each which still exists today is also given.

As you can see from Table 9, most of the nine have not seriously

TABLE 9 — THE EFFECTIVELY RADIOACTIVE LONG-LIVED NINE

Nuclide	Half-life (in eons)	Fraction of original remaining today
Lanthanum-138	110	0.97
Samarium-147	106	0.97
Rhenium-187	70	0.95
Rubidium-87	47	0.93
Lutetium-176	21	0.86
Thorium-232	13.9	0.79
Uranium-238	4.51	0.50
Potassium-40	1.30	0.086
Uranium-235	0.713	0.012

diminished in the course of the Earth's history. Even thorium-232, sixth on the list, is still present in four-fifths of its original quantity. The only really serious diminutions are those of the last three nuclides on the list. We have only one half remaining of the original uranium-238 with which Earth was supplied, only one twelfth the potassium-40, and only one eightieth the original uranium-235.

Of the nine effectively radioactive long-lived nuclides, consider potassium-40. It is the least massive nuclide to demonstrate long-lived radioactivity. In general, nuclides with small mass are more common in the Universe than those with large mass, so we might suspect that potassium-40 is the most common or, at the very least, one of the most common of the long-lived nuclides.

We can check this. The relative quantity of the elements in the

Earth's crust is known in a very rough way, but we can use that as the starting point, for what it's worth. Then, the relative quantity of a particular nuclide in a given element is also known, and we can use that.

For instance, the amount of potassium in the soil is estimated as 25,900 parts per million. Since potassium-40 makes up only 0.0119 per cent of all the potassium atoms, we can say that the potassium-40 content of the soil is about 3.08 parts per million. In this way we can prepare Table 10, which gives the values (very rough ones) for the quantity present in the Earth's crust of each of the twenty-one long-lived nuclides listed in Table 8.

TABLE 10 — THE RADIOACTIVE NUCLIDES IN EARTH'S CRUST

Nuclide	Quantity in Earth's crust (in parts per million)
Calcium-48	67.2
Rubidium-87	33.6
Thorium-232	10
Neodymium-144	5.6
Cerium-142	5.1
Potassium-40	3.08
Uranium-238	2
Samarium-147	1
Samarium-149	0.9
Samarium-148	0.8
Vanadium-50	0.26
Lead-204	0.22
Hafnium-174	0.1
Indium-115	0.1
Lanthanum-138	0.02
Lutetium-176	0.02
Uranium-235	0.015
Gadolinium-152	0.01
Rhenium-187	0.0006
Platinum-190	0.00005
Platinum-192	0.0000006

It would appear, from Table 10, that calcium-48 is by far the most common radioactive nuclide on Earth. There is about as much calcium-48 in the Earth's crust as there is all other radioactive nuclides combined.

Yet, really, that is not very impressive. Calcium-48 has such a long half-life (twenty million eons) that the number of its atoms breaking down per second cannot be very impressive. It would be quite easy for nuclides present in lesser concentration but with a *far* lesser half-life to outdo it enormously in this respect. And it is, after all, the number of breakdowns per second that is a more natural measure of the importance of a radioactive nuclide than the mere accumulation of inactive mass.

Why not convert Table 10 into one that measures the number of atomic breakdowns per second for each nuclide? To avoid astronomical figures, I won't try to deal with the total number of breakdowns per second over all the Earth's crust, but will give the relative numbers by setting the value for uranium-238 arbitrarily at 1. The result is Table 11, in which, we see, only six of the nuclides have figures large enough to be worth noting.

We can see from Table 11 that potassium-40 does indeed dominate the field. If we consider all the long-lived radioactive nuclides on Earth, it turns out that more than four fifths of all the breakdowns taking place each second are breakdowns of potassium-40.

That is the situation as it is *now*. What about the past? What was the situation at the time of the Earth's beginning 4.6 eons ago?

Each of the radioactive nuclides listed in Table 11 has shrunk in quantity since Earth's origin at least several per cent. Those of the entire group of twenty-one which are *not* listed in Table 11 are so weakly radioactive and have therefore shrunk so little in quantity that we can, without serious error, consider their contribution both trifling now and equally trifling at the time of the Earth's formation.

Taking into account, then, only those nuclides listed in Table 11, we can calculate the amount present of each in the Earth's crust at the start and reach the figures presented in Table 12. In Table 13 we have the relative number of breakdowns per second for those nuclides as they took place at the time of Earth's formation.

From the standpoint of the elements, the biggest difference in the Earth's crust at the time of formation as compared with the

TABLE 11 — NUCLIDE BREAKDOWNS

Nuclide	Breakdowns per second (uranium-238=1)
Potassium-40	31
Rubidium-87	4.3
Thorium-232	1.6
Uranium-238	1.0
Uranium-235	0.05
Samarium-147	0.03
All others	0.004

TABLE 12 — THE RADIOACTIVE NUCLIDES IN EARTH'S CRUST AT THE BEGINNING

Nuclide	Quantity in Earth's crust at the beginning (in parts per million)
Potassium-40	36.8
Rubidium-87	36.0
Thorium-232	12.7
Uranium-238	4.0
Samarium-147	1.0
Uranium-235	0.12

TABLE 13 — NUCLIDE BREAKDOWNS IN EARTH'S CRUST AT THE BEGINNING

Nuclide	Breakdowns per second at the beginning (uranium-238=1)
Potassium-40	180
Rubidium-87	2.3
Uranium-235	2.0
Thorium-232	1.0
Uranium-238	1.0
Samarium-147	0.02
All others	0.002

present is in the uranium content. The only nuclides found in uranium as it occurs naturally are uranium-238 and uranium-235, both of which have comparatively high levels of radioactive breakdown. As a result, Earth's crust, at the time of its formation, was twice as rich in the element uranium as it is now.

What's more, the ratio of the nuclide content was different. At the present moment, uranium is 99.28 per cent uranium-238 and only 0.72 per cent uranium-235. Since uranium-235 has a half-life less than a sixth that of uranium-238, it has been disappearing considerably more quickly. At the time of Earth's formation, the element was something like 97 per cent uranium-238 and 3 per cent uranium-235.

Although of the long-lived nuclides, potassium-40 disappeared more rapidly than did any of the others but uranium-235, this did not significantly affect the total quantity of potassium in the Earth's crust. Unlike uranium, potassium is not composed of radioactive nuclides only. Indeed, 99.99 per cent of potassium is made up of the two stable nuclides, potassium-39 and potassium-41. Potassium-40 makes up only 0.0119 per cent of the element.

At the start, potassium-40 was nearly twelve times as abundant as it is now, but even then it was still only about 0.13 per cent of all the potassium. That larger percentage is still so small that it doesn't significantly alter the total quantity.

Notice, however, that at the time of the Earth's formation, the number of breakdowns of potassium-40 was a truly overwhelming majority of the total number of breakdowns of the long-lived radioactive nuclides. Over 96 per cent of all atomic breakdowns of long-lived radioactive isotopes on Earth involved potassium-40.

The vast predominance of potassium-40 radioactivity in the Earth's original crust and the lesser but still considerable predominance of the nuclide today raise two questions:

1. Why is so little attention given to potassium-40 and so much to uranium-238 and thorium-232, when one deals with such matters as, for instance, the effect of radioactivity in heating up Earth's interior?

2. The six nuclides listed in Table 11 were the first to be discovered to be radioactive, which is not surprising since they experience far more breakdowns than all other long-lived radioactive nuclides put together. Yet it was the radioactive properties of

uranium and thorium that were first discovered, in the late 1890s, whereas the radioactive properties of potassium and rubidium were not detected till 1906. Why the ten-year delay in detecting the far greater number of breakdowns of the latter?

I'll go into that in the next chapter.

13.   A Particular Matter

My beautiful, blonde-haired, blue-eyed daughter, who has been figuring in the various essays I write since she was a preschooler is now (as I write this) a new freshman in college. (Oddly enough, I myself haven't aged a day.)

It seems that she had not been at college for long when word reached her that a young man on another floor in the dormitory was dying to meet her, presumably, she was given to understand by an excited informant, to feast his eyes on her beauty. After some ladylike show of reluctance, Robyn agreed to allow the meeting.

In came the young man, eyes wide. "Tell me," he said, choking a bit, "are you *really* the daughter of Isaac Asimov?"

My daughter called me afterward, of course, and said to me, severely, "The worst of it was that he seemed to know everything about our private lives because he's memorized everything you've written."

Well, touché. I *do* have a personal writing style and I *do* suffer from a lack of reticence, and sometimes I feel defensive about it.

Yet ask yourself, if I had started this essay with: "The two chief methods by which a radioactive nucleus can break down involve, one, the emission of an alpha particle, and, two, the emission of a beta particle," would you have plunged in quite as readily as you would with a beginning like: "My beautiful, blonde-haired, blue-eyed daughter, who has been figuring in the various essays I write since she was a preschooler—"?

So I hope Robyn will understand that the sacrifice of a little bit of her private life may be essential to the important function of paying those college bills plus ancillary expenses. (It's those ancillaries that get you every time.)

And now that that's taken care of—

The two chief methods by which a radioactive nucleus can break down involve, one, the emission of an alpha particle, and, two, the emission of a beta particle.

Of the twenty-one long-lived radioactive nuclides I listed in the previous chapter, eight break down by means of beta-particle emission. Let's see what this means.

A beta particle is a speeding electron, and an electron carries a negative electric charge. It is produced when, inside the nucleus, a neutron (uncharged) is converted into a proton (positive) and an electron (negative). The proton remains in the nucleus and the electron goes flying out of it at great speeds.

The mass of the nucleus (its "mass number") is taken as equal to the total number of protons and neutrons within the nucleus. Since this remains unchanged by any conversion of neutrons to protons, or vice versa, the mass number remains unchanged when a beta particle is emitted.

The total number of protons is, however, increased by one, when a neutron turns into a proton. This means that the atomic number, which is equal to the number of protons, is increased by one. Since a change in atomic number means a change in the nature of the element, a nuclide that emits a beta particle becomes a nuclide of another element, the one a single unit higher up in the atomic number scale.

Let's take an example. Indium has an atomic number of 49. That means that every indium atom has 49 protons in its nucleus. Indium-115 has a mass number of 115 (the number attached to a nuclide is always its mass number), so that it possesses a total of 115 protons and neutrons. If 49 of this total are protons, then 66 are neutrons.

When indium-115 emits a beta particle, one neutron turns to a proton and the total present becomes 50 protons and 65 neutrons. The element with an atomic number of 50 is tin, so by emitting a beta particle, indium-115 becomes the stable nuclide tin-115.

It is possible, of course, for a nuclide to break down, by way of beta-particle emission to a nuclide that is also unstable and breaks down further. For instance, calcium-48 (20 protons+28 neutrons) emits a beta particle and becomes scandium-48 (21 protons+27 neutrons). Scandium-48 is unstable and very short-lived and quickly emits a beta particle of its own to become the stable nuclide titanium-48 (22 protons+26 neutrons).

It is also possible for the reverse action to take place within a nucleus, for a proton to become a neutron. This can happen when a proton absorbs one of the electrons that is circling the nucleus and thus becomes a neutron.

In the case of such electron-capture, the mass number of the nucleus remains unchanged, but the atomic number goes *down* by one. Thus vanadium-50 (23 protons+27 neutrons) may capture an electron to become titanium-50 (22 protons+28 neutrons).

None of the long-lived nuclides break down by electron-capture exclusively. Some of those that do (three, all told) also break down by beta-particle emission. Five break down by beta-particle emission only. The details are presented in Table 14, where each nuclide has its proton (p) and neutron (n) content included.

The next question is this: How many beta particles are emitted by each of these eight nuclides?

If you compare equal numbers of atoms of different nuclides, you would find that the number of beta particles emitted per second is inversely proportional to the half-life. Thus, the half-life of indium-115 is 600,000 eons (where an eon, as I explained in the last chapter, is equal to one billion years) while that of lutetium-176 is only 21 eons. This means that lutetium-176 must be breaking down some thirty thousand times faster than indium-115. Therefore, for every beta particle emitted by a particular large number of indium-115 nuclides, thirty thousand beta particles are being emitted by the same large number of lutetium-176 nuclides.

But equal numbers of the two do not exist in the Earth's crust. The weight of lutetium in Earth's crust is roughly eight times that of indium. Lutetium-176, however, makes up only about 2.6 per cent of all the lutetium there is, while indium-115 makes up about 95.8 per cent of all the indium. Therefore indium-115 is about five times as common in the crust as lutetium-176 is, at least, in terms of weight.

The total number of nuclides in a given weight is, however, inversely proportional to the atomic weight. In one gram of indium-115 there are 176/115, or 1.53 , as many nuclides as in one gram of

TABLE 14 — THE BETA-PRODUCING NUCLIDES

Parent nuclide	Nature of breakdown	Daughter nuclide
Potassium-40 $(19p+21n)$	Beta particle Electron-capture	Calcium-40 $(20p+20n)$ Argon-40 $(18p+22n)$
Calcium-48 $(20p+28n)$	2 beta particles	Titanium-48 $(22p+26n)$
Vanadium-50 $(23p+27n)$	Beta particle Electron-capture	Chromium-50 $(24p+26n)$ Titanium-50 $(22p+28n)$
Rubidium-87 $(37p+50n)$	Beta particle	Strontium-87 $(38p+49n)$
Indium-115 $(49p+66n)$	Beta particle	Tin-115 $(50p+65n)$
Lanthanum-138 $(57p+81n)$	Beta particle Electron-capture	Cerium-138 $(58p+80n)$ Barium-138 $(56p+82n)$
Lutetium-176 $(71p+105n)$	Beta particle	Hafnium-176 $(72p+104n)$
Rhenium-187 $(75p+112n)$	Beta particle	Osmium-187 $(76p+111n)$

lutetium-176. Taking all this into account, there are about seven times as many indium-115 nuclides as there are lutetium-176 nuclides.

Allowing for the difference in quantity of the two nuclides in the Earth's crust, then, it turns out that for every beta particle emitted by indium-115, about four thousand are emitted by lutetium-176.

Two other points must be taken into account. In the conversion of calcium-48 to titanium-48, *two* beta particles are emitted per atom.

Then, too, in the case of three nuclides, potassium-40, vanadium-50, and lanthanum-138, some individual nuclides emit a beta particle and some capture an electron. In the latter case, no particle is emitted.

Potassium-40 is sufficiently radioactive to make it possible to determine the proportion of individual atoms taking either route. It turns

out that 89 per cent of them indulge in beta-particle emission and
11 per cent in electron-capture. That can be allowed for. In the
case of the other two nuclides, radioactivity is so weak that the
fraction following either route has not been determined (at least
as far as I know). In their cases, we will assume, for simplicity's
sake, that it is all beta-particle emission.

Keeping all this in mind, we can calculate the relative production
of beta particles within the Earth's crust by these eight radioactive
nuclides. The results (which can only be considered approximate
since neither the half-lives nor the relative contents of the Earth's
crust are known with great accuracy) are presented in Table 15.

As you see from that table, the six least active beta-particle
producers, taken together, account for only a little over 1/10,000th
of the total. This is so small it can safely be ignored and we can
say that of these nuclides, the only important sources of beta particles

TABLE 15 — BETA-PARTICLE PRODUCTION

Nuclide	Number of beta particles produced (indium-115=1)
Indium-115	1
Vanadium-50	6
Rhenium-187	33
Calcium-48	96
Lanthanum-138	730
Lutetium-176	4,000
Rubidium-87	5,600,000
Potassium-40	36,000,000

are potassium-40 and rubidium-87, with the former supplying about
seven eighths of the total and the latter the remaining one eighth.

But what about radioactive breakdown by the emission of alpha
particles?

Alpha particles consist of a firm association of two protons and
two neutrons. When a nuclide emits an alpha particle, it loses those
two protons and two neutrons. Its mass number therefore goes down

by four units, and its atomic number, dependent on protons only, goes down by two units.

Let's take an example. Lead has an atomic number of 82 so that all its atoms have 82 protons in their nuclei. The nuclide lead-204 has a total of 204 protons and neutrons present in the nucleus and since 82 of them are protons, 122 of them must be neutrons.

Lead-204 emits an alpha particle. Its atomic weight goes down to 200, and its atomic number goes down to 80. Its nucleus now contains 80 protons and 120 neutrons. Since the element with an atomic number of 80 is mercury, the nuclide resulting from the alpha-particle emission of lead-204 is the stable mercury-200.

Of the twenty-one long-lived radioactive nuclides in the Earth's crust, thirteen break down by way of alpha-particle emission. Of these thirteen, three are thorium-232, uranium-238, and uranium-235. These three I will leave aside for the moment and concentrate on the remaining ten, which are listed in Table 16.

(Note that in two cases, that of gadolinium-152 and samarium-148, the daughter nuclide is itself one of the long-lived varieties contained in the table and is therefore marked with an asterisk. We can ignore that slight complication, however.)

We can next calculate the relative numbers of alpha particles produced in the Earth's crust by these nuclides. In order that these be compared directly with the beta-particle figures given in Table 15, I have calculated the figures for alpha-particle production to the same arbitrary base which sets indium-115-particle production equal to 1. The results are given in Table 17.

As you can see, the alpha-particle production by these ten nuclides taken together is rather insignificant compared to the beta-particle production of potassium-40 and rubidium-87. Even samarium-147, which produces about 99.3 per cent of all the alpha particles for which these ten nuclides are responsible, produces only about 1/120 as many particles as rubidium-87 does.

But now let's get on to the three alpha-particle producers I have left out: thorium-232, uranium-235, and uranium-238.

When each of these produces an alpha particle, it breaks down to a nuclide that is *not* stable. This is also true, to be sure, of gadolinium-152, which breaks down to samarium-148, which in turn breaks down to neodymium-144, which in turn breaks down to the

TABLE 16 — THE ALPHA-PRODUCING NUCLIDES

Parent nuclide	Nature of breakdown	Daughter nuclide
Cerium-142 $(58p+84n)$	Alpha particle	Barium-138 $(56p+82n)$
Neodymium-144 $(60p+84n)$	Alpha particle	Cerium-140 $(58p+82n)$
Samarium-147 $(62p+85n)$	Alpha particle	Neodymium-143 $(60p+83n)$
Samarium-148 $(62p+86n)$	Alpha particle	Neodymium-144 $(60p+84n)$*
Samarium-149 $(62p+87n)$	Alpha particle	Neodymium-145 $(60p+85n)$
Gadolinium-152 $(64p+88n)$	Alpha particle	Samarium-148 $(62p+86n)$*
Hafnium-174 $(72p+102n)$	Alpha particle	Ytterbium-170 $(70p+100n)$
Platinum-190 $(78p+112n)$	Alpha particle	Osmium-186 $(76p+110n)$
Platinum-192 $(78p+114n)$	Alpha particle	Osmium-188 $(76p+112n)$
Lead-204 $(82p+122n)$	Alpha particle	Mercury-200 $(80p+120n)$

stable cerium-140. Each of these nuclides, however, is long-lived, so they can be considered separately.

Then, too, we have the case of calcium-48, which produces a beta particle to become scandium-48, which in turn produces a beta particle to become the stable nuclide, titanium-48. Here, where calcium-48 has a half-life of 20,000,000 eons, scandium-48 has a half-life of only 44 *hours*. The half-life for the production of the second beta particle is so much enormously shorter than for the first that we are perfectly safe in saying that calcium-48 gives up two beta particles altogether and working out our calculations on that basis.

We must do the same for thorium-232, uranium-235, and uranium-

* Itself a long-lived alpha-producer.

TABLE 17 — ALPHA-PARTICLE PRODUCTION

Nuclide	Number of alpha particles produced (indium-115=1)
Lead-204	0.00005
Platinum-192	0.00014
Platinum-190	0.0026
Hafnium-174	0.008
Gadolinium-152	0.5
Neodymium-144	5.1
Cerium-142	5.7
Samarium-149	11
Samarium-148	310
Samarium-147	46,500

238, but that is not as easy as it is with calcium-48. In the latter case only one short-lived nuclide is produced before stability is reached; in the case of the uranium and thorium nuclides about a dozen are. In the case of calcium-48 only beta particles are involved; in the case of the uranium and thorium nuclides both alpha particles and beta particles are.

Thorium-232, for instance, gives off an alpha particle to become the short-lived radium-228, which gives off a beta particle to become the short-lived actinium-228, which gives off a beta particle to become the short-lived—and so on and so on.

The over-all result is that we must consider the thorium and uranium nuclides as giving off a number of alpha particles and beta particles for each individual nuclide that breaks down before reaching stability (in each case) with one of the lead nuclides, as shown in Table 18.

Once we have the data given in Table 18 it is possible to calculate the number of alpha particles and beta particles produced by the thorium and uranium nuclides on the usual indium-115=1 standard. This is given in Table 19.

From Tables 15, 17, and 19, we can choose those nuclides which produce more than a million particles by the indium-115=1 standard and prepare Table 20.

TABLE 18 — THORIUM AND URANIUM

Parent nuclide	Nature of breakdown	Daughter nuclide
Thorium-232 $(90p+142n)$	6 alpha particles $(12p+12n)$ 4 beta particles	Lead-208 $(82p+126n)$
Uranium-235 $(92p+143n)$	7 alpha particles $(14p+14n)$ 4 beta particles	Lead-207 $(82p+125n)$
Uranium-238 $(92p+146n)$	8 alpha particles $(16p+16n)$ 6 beta particles	Lead-206 $(82p+124n)$

TABLE 19 — THORIUM AND URANIUM PARTICLE PRODUCTION

Parent nuclide	Particles produced $(indium-115=1)$	
	Alpha particles	Beta particles
Uranium-235	410,000	240,000
Uranium-238	10,400,000	7,800,000
Thorium-232	12,600,000	8,400,000

TABLE 20 — THE MAJOR PARTICLE-PRODUCERS

Nuclide	Particles produced $(indium-115=1)$		
	Alpha particles	Beta particles	Total particles
Rubidium-87	———	5,600,000	5,600,000
Uranium-238	10,400,000	7,800,000	18,200,000
Thorium-232	12,600,000	8,400,000	21,000,000
Potassium-40	———	36,000,000	36,000,000

Now I am ready to consider the question which I had raised at the end of the last chapter. I had shown there that by far the large majority of the nuclide breakdowns taking place in the Earth's crust involved potassium-40 and rubidium-87. Why, then, was the radioactivity of uranium and thorium discovered ten years *before* the radioactivity of potassium and rubidium was?

Now, though, we have taken into account not only the breakdown of the nuclides, but also the number of particles that are produced as a result of the breakdown. Because the breakdown of uranium-238 and thorium-232 result in the production of about a dozen particles per nuclide breaking down, the disproportion between potassium and rubidium on the one hand and uranium and thorium on the other is reduced considerably.

If we consider the total number of particles produced by the natural radioactivity of the Earth's crust, rather than the total number of nuclide breakdowns (and it is, after all, the particles that we detect), it turns out that potassium-40 produces about 45 per cent of them, thorium-232 about 26 per cent, uranium-238 about 22 per cent, and rubidium-87 about 7 per cent. All other long-lived radioactive nuclides in the Earth's crust contribute less than 1 per cent of the particles produced.

And yet even by counting the multiple-particle production of uranium and thorium, the division is still roughly fifty-fifty for uranium and thorium on the one hand, and potassium and rubidium on the other. The reason for the ten-year lapse in the discovery of the radioactivity of the latter nuclides is still to be explained.

The answer lies in the fact that we must not consider *total* particles. Alpha particles and beta particles are not detected with equal ease. The ease of detection depends on the energy of the particles, which in turn depends upon the mass and upon the square of the velocity. Alpha particles are some seven thousand times as massive as beta particles, while beta particles speed along with some ten times the speed of the alpha particles. Combining the two we see that the average alpha particle has some seventy times the energy of the average beta particle and are correspondingly easier to detect.

Well, then, uranium and thorium between them produce 23,000,-000 of the comparatively easily detected alpha particles while potassium and rubidium between them produce exactly none! *That* is the difference that accounts for the ten-year lapse before the discovery of the radioactivity of the latter.

Let's look at the matter from a different angle. There are six known nuclides with mass numbers of 87, but the actual mass of each is not *quite* 87 and is, in each case, different, as you can see in Table 21.

In that table you can see that strontium-87 has the lowest mass in the series. It lies at the bottom of the mass-valley and is the one nuclide of that mass number that is stable. The other nuclides are all radioactive, and the larger their mass, the greater the push to undergo a radioactive change and the shorter the half-life of that change. (The mass-excess is, to a large extent, liberated as energy.)

TABLE 21 — NUCLIDES OF MASS 87

Nuclide	Mass	Mass-excess (millionths)	Half-life
Bromine-87 ($35p+52n$)	86.949	143	55.6 seconds
Krypton-87 ($36p+51n$)	86.9412	53	78 minutes
Rubidium-87 ($37p+50n$)	86.93687	3.3	46 eons
Strontium-87 ($38p+49n$)	86.93658	0.0	stable
Yttrium-87 ($39p+48n$)	86.9384	21	80 hours
Zirconium-87 ($40p+47n$)	86.9422	65	94 minutes

TABLE 22 — NUCLIDES OF MASS 40

Nuclide	Mass	Mass-excess (millionths)	Half-life
Chlorine-40 ($17p+23n$)	c. 39.985	c. 200	1.4 minutes
Argon-40 ($18p+22n$)	39.97505	0.0	stable
Potassium-40 ($19p+21n$)	39.97665	40	1.3 eons
Calcium-40 ($20p+20n$)	39.97523	4.5	stable
Scandium-40 ($21p+19n$)	39.9902	380	0.22 seconds

The mass-excess is given in Table 21 for each nuclide in terms of millionths of the mass number, and as you can see, the half-life is very sensitively dependent on that relative mass-excess. Rubidium-87, which has a very small relative mass-excess, has a half-life of 46 billion years, while yttrium-87, with a relative mass-excess less than seven times as large as that of rubidium-87, has a half-life of only 80 hours.

We can do the same for the known nuclides of mass 40, and these are given in Table 22. Here you can see there are two valleys. Argon-40 has a lesser mass than the nuclide before and after and the same is true of calcium-40. Both these nuclides are therefore stable. As for potassium-40, that is only slightly higher in mass than the nuclide before and after so, although radioactive, it is long-lived.

Again, since the nuclide before potassium-40 and the one after are each lower in mass than potassium-40 is, potassium-40 can move in either direction. It can emit a beta particle and become calcium-40 or absorb an electron and become argon-40.

For reasons involving the proton-neutron arrangement within the nucleus, potassium-40 has a greater tendency to emit a beta particle and become calcium-40, although calcium-40 is slightly the more massive of the two stable nuclides of mass 40. In that direction, the mass-excess of potassium-40 is 36 and that is the number we should use.

To handle uranium-238 and thorium-232 requires an additional complication, for there we have to count in the mass of alpha particles, which, like the atomic nuclei generally, contain both protons and neutrons.

Thus uranium-238 gives off an alpha particle to become thorium-234. The mass of the uranium-238 nucleus is 238.12522, while the mass of the thorium-234 and the alpha particle taken together is 238.12065. The mass-excess of the uranium-238 nuclide over its immediate product in millionths of the mass number is 20. If we do the same for thorium-232, then the excess of that nuclide over its immediate product in millionths of the mass number is 19.

But we can't stop with the immediate product of uranium-238 and thorium-232, which are thorium-234, and radium-228 respectively. Each of these continues to break down, until lead-208 and lead-206, respectively, are produced. In the process, the breakdown of uranium-238 and its daughter nuclides produces eight alpha particles altogether, while the breakdown of thorium-232 and its daughter nuclides produces six alpha particles altogether.

The mass-excess of uranium-238 over lead-206 and eight alpha particles, in millionths of the original mass number, is 230; that of thorium-232 over lead-208 and six alpha particles is 200.

We can summarize this in Table 23.

Remember that the mass-excess controls, to a large extent, the energy of the breakdown; that the energy of the breakdown controls, to a large extent, the energy of the particles produced; that the energy of the particles produced controls, to a large extent, the ease with which they are detected.

In that case, if uranium-238 and thorium-232 broke down as they did but produced a stable product after the loss of a single alpha particle, their breakdowns would, on the whole, be less energetic than potassium-40 and more energetic than rubidium-87. To be sure, uranium-238 and thorium-232 would still have the advantage of shooting off an alpha particle, in comparison to the potassium-40's less easily detected beta particle. That, however, might not be enough to guarantee primacy of discovery to uranium and thorium. It might well have been, in that case, that potassium would have been the element whose radioactivity was first to be detected.

However, the daughter nuclides of uranium-238 and thorium-232 slide down the radioactive hill, pushed by much larger mass-excesses than those involving the parent nuclide. The daughter nuclides

TABLE 23 — MASS-EXCESSES

Nuclide	Mass-excess (millionths)	Half-life
Rubidium-87 (no alphas)	3.3	46 eons
Thorium-232 (1 alpha)	19	14 eons
Uranium-238 (1 alpha)	20	4.5 eons
Potassium-40 (no alphas)	36	1.3 eons
Thorium-232+daughter nuclides (6 alphas)	200	down to fractions of a second
Uranium-238+daughter nuclides (8 alphas)	230	down to fractions of a second

have much shorter half-lives, therefore, down to fractions of a second in some cases, and produce particles of much greater energies.

These daughter nuclides are invariably associated with uranium and thorium as found in nature, and it was their energetic particles, *theirs*, that were first detected and (naturally) attributed to the uranium and thorium that contained them.

From Table 23 we see that the relative mass-excess is 230 for uranium-238 *plus its daughter nuclides*, 200 for thorium-232 *plus its daughter nuclides*, 36 for potassium-40, and 3.3 for rubidium-37. And that order—uranium, thorium, potassium, and rubidium—is exactly the order of the first four elements in which radioactivity was discovered.

No mystery at all.

What's more, you can also see why it is that when geologists talk about the warming of the Earth through radioactive breakdown in the crust, they are concerned almost entirely with uranium-238 and thorium-232 and scarcely at all with any other nuclide. —What produces the warming effect are those daughter nuclides—those beautiful, blonde-haired, bl——

No, that's a different daughter.

14. At Closest Range

Confession, they say, is good for the soul, so, if you don't mind, I'm going to confess something.

Before I began this series of essays for good old *F & SF* I wrote essays of the same sort for another science fiction magazine (*mea culpa*). Then, a year before my *F & SF* series began, I gave birth to a book entitled *Only a Trillion* (Abelard-Schuman, 1957), in which these prehistoric non-*F & SF* essays were collected.

The trouble was that my early essays were written before I had developed the "article-style" to which I am now accustomed (see Introduction), so that, as the years passed, I have come to feel a little restless about *Only a Trillion* and to deny it, in my mind, a position of equality with the collections of my *F & SF* essays that have so far appeared in book form.

I keep sighing and wishing I could recast the essays into a form that would suit me today.

This is especially true of the first two chapters of the book. It seems that the editor at Abelard-Schuman rejected one of the essays I had originally included in the manuscript, feeling it wasn't sufficiently interesting. (She was quite wrong, of course, and I later expanded the reject into an entire successful book.) Because of the rejection, however, I had to supply something for the hole left behind. I therefore very quickly adapted several essays I had earlier written for a scholarly periodical entitled *The Journal of Chemical Education* and that was what made up the first two chapters.

Now over sixteen years have passed, and though I have manfully resisted recasting those chapters of the book which had actually appeared in *Astounding* (oops!), I can no longer resist redoing those first chapters which, after all, have never been presented to my real audience.

Indeed, I have already done so in part. The previous two chapters have retold, in more detail and in updated fashion, the subject matter of that first chapter, and now I intend to take up the subject matter of the second.

The human body is made up of a variety of elements, all of them very common and unglamorous. This comes as a shock to some people, who think that something as mysterious and subtle as life should be found only in objects containing some rare and magic ingredients.

This is unlikely, though, on the face of it. Life is so common a phenomenon on Earth that it couldn't possibly depend on anything rare. How many automobiles could we build if the engines had to be made of solid gold?

But what the heck, it isn't the elements that make life what it is, but the arrangement of the atoms of those elements into molecules that are, in some cases, extraordinarily complex, and the further arrangement of those molecules into systems that are still more extraordinarily complex.

It comes about, therefore, that although we know a great deal about the elements that make up the human body, we know considerably less about the more complicated molecules into which those elements are organized, and, even today, virtually nothing about the systems into which those molecules are organized.

Adopting the admirable practice, however, of talking about something I know about, I will consider the elements and see if I can extract something interesting out of them.

The human body, to begin with, is mostly water. In fact, if we were actually to count all the molecules in the human body, it would turn out that 98 per cent of them would be water molecules. However, water molecules are very small (each being made up of but three atoms, two of hydrogen and one of oxygen) and the other molecules in living tissue are considerably larger—some being composed of millions of atoms. As a result, all those water molecules make up only about 65 per cent, by weight, of the human body. (This varies from tissue to tissue. Our lordly brain, the crown and summit of Homo sapiens, is 84 per cent water, by weight.)

Of the two types of atoms making up the water molecule, the

oxygen atom is sixteen times as massive as the hydrogen atom. Since there are two hydrogen atoms for every oxygen in the water molecule, the weight ratio is eight to one in favor of oxygen. If we add in the oxygen and hydrogen present in living tissue in molecules other than water, it turns out that, by weight, we are about 65 per cent oxygen and 10½ per cent hydrogen. Three fourths of the weight of the human body is made up of those two elements alone.

The chemical recipe for the human body is given in Table 24. As you can see there, about 96.5 per cent of the weight of the human body is made up of four elements: oxygen, carbon, hydrogen, and nitrogen, in that order. Considering that the atmosphere is four fifths nitrogen and one fifth oxygen, that water is made up of oxygen and hydrogen, and that coal is almost entirely carbon, we see how unimpressive the elemental composition of the human body is.

(Of course, we must not underestimate the minor constituents of the human body. Included in the "everything else" listed in Table 24 are manganese, iodine, copper, cobalt, zinc, molybdenum, chromium, selenium, and vanadium, each of which, while present in relatively tiny quantities, makes up essential parts of essential molecules. Without the small quantities of each of these "trace elements" in our body we would die just as surely as if we lacked oxygen or carbon.)

But we can list the ingredients of the body a little more fundamentally than by element. Not all the atoms of a particular element are alike. They all have the same number of protons in their nucleus, but they may differ in the number of neutrons.

For instance, all oxygen atoms have eight protons in their nucleus, but some may have eight neutrons, some nine, and some ten. The total number of nuclear particles in the nucleus of an oxygen atom can therefore be sixteen, seventeen, or eighteen. We can speak, then, of three different oxygen nuclides: oxygen-16, oxygen-17, and oxygen-18.

In the same way, hydrogen, carbon, and nitrogen are each made up of two nuclides: hydrogen-1 and hydrogen-2, carbon-12 and carbon-13, and nitrogen-14 and nitrogen-15.

TABLE 24 — ELEMENTS IN THE HUMAN BODY

Element	Weight (parts per hundred)
Oxygen	65
Carbon	18
Hydrogen	10.5
Nitrogen	3.0
Calcium	1.5
Phosphorus	0.9
Potassium	0.4
Sulfur	0.3
Chlorine	0.15
Sodium	0.15
Magnesium	0.05
Iron	0.006
Everything else	0.004

As it happens, in the case of each of these four elements, one nuclide makes up an overwhelming proportion of all its atoms. Oxygen-16 makes up 99.76 per cent of all oxygen atoms, carbon-12 makes up 98.89 per cent of all carbon atoms, hydrogen-1 makes up 99.985 per cent of all hydrogen atoms, and nitrogen-14 makes up 99.635 per cent of all nitrogen atoms.

We can therefore go farther than to say that some 96.5 per cent of the weight of the human body is made up of four elements: oxygen, carbon, hydrogen, and nitrogen. We can say that some 96.2 per cent of the weight of the human body is made up of four *nuclides:* oxygen-16, carbon-12, hydrogen-1, and nitrogen-14.

Suppose, then, we prepare a list of the nuclides in the body, and list them in order of weight. The result, which includes only those nuclides present in greater quantity than hydrogen-2 (and omits quite a few that are present in lesser quantity), is given in Table 25.

There's more chance for glamor here. Back in 1931, when hydrogen-2 was first discovered, it was popularly called "heavy hydrogen" and for a while it seemed an exciting substance indeed. In combination with oxygen, it made up "heavy water" and there were a number of science fiction stories written that dealt with the mysterious properties of heavy water.

I suspect that the lay public got the notion that heavy water was present nowhere but in the glamorous recesses of the laboratories of some mad scientists. Actually, of course, it was present wherever ordinary water was present; in fact, wherever ordinary hydrogen atoms were present. This meant that living tissue contained heavy hydrogen and heavy water. If we consider a human body weighing

TABLE 25 — NUCLIDES IN THE HUMAN BODY

Nuclides	Weight (parts per thousand)
Oxygen-16	
Carbon-12	962
Hydrogen-1	
Nitrogen-14	
Calcium-40	14.5
Phosphorus-31	9.0
Potassium-39	3.7
Sulfur-32	3.0
Carbon-13	2.0
Sodium-23	1.5
Oxygen-18	1.3
Chlorine-35	1.1
Chlorine-37	0.4
Calcium-44	0.3
Oxygen-17	0.23
Sulfur-34	0.12
Nitrogen-15	0.11
Potassium-41	0.03
Iron-56	0.03
Magnesium-24	0.025
Hydrogen-2	0.016

75 kilograms, its content of hydrogen-2 would be 1.2 grams; enough to make up 10 grams of heavy water if combined with oxygen, which most of it would be.

Imagine—10 grams of heavy water in your body. Wow! —Except that heavy water isn't nearly as mysterious or glamorous as the science fiction of the 1930s had thought it might be.

If hydrogen-2 isn't glamorous, are there any other nuclides that are? Of course. The direction in which to look is obvious. In the previous two chapters I dealt with long-lived radioactive nuclides, and we might ask ourselves if any of those nuclides are present in the human body.

Some surely are, since we live surrounded by all the different elements of the world and a little bit of everything must surely find its way into our body in the way of impurities, at the very least. If we could go over our bodies atom by atom, we would find a few atoms of gold, a few of platinum, a few of uranium, and so on. In short, we would find a few of every variety that exists in our surroundings, including some radioactive ones.

Let's, however, eliminate accidental contamination, which is hard to measure and, in any case, must be highly variable.

Let's ask, instead, whether any of the elements known to be essential to life, and therefore present *of necessity* in the human body, possess radioactive nuclides. If we go over the list of long-lived radioactive nuclides that I dealt with in the last two chapters, we find that most of them occur in elements that are not (as far as we know) essential to human life—neodymium, samarium, gadolinium, hafnium, and so on.

Two long-lived radioactive nuclides, however, are members of elements that have long been known to be essential to life and each of these is and *must be* (unless food is painfully prepared to eliminate those nuclides—something that has never been done, as far as I know) present in the human body. These two are calcium-48 and potassium-40.

Vanadium has recently been reported as an essential trace element, and if that is so, vanadium-50 should be included as a third long-lived radioactive nuclide inevitably present in living tissue.

If vanadium is indeed present as a trace element in the body, it would be present in perhaps 0.1 parts per million by weight. Most of the vanadium atoms are, however, vanadium-51, which is stable. Only 0.24 per cent of vanadium atoms are vanadium-50.

It follows that vanadium-50 is present in the body in only 0.0002 parts per million.

Calcium-48 is an even rarer component of its element. Only 0.185 per cent of all the calcium atoms that exist are calcium-48. Calcium, however, is present in the body to a much greater extent than vanadium is, so calcium-48 is present to the extent of 28 parts per million. Potassium-40 is present in its element in still smaller proportion—0.0119 per cent—and potassium is between the other two in body content. Potassium-40 is present in the body in 0.5 parts per million.

We can work this out for a 75-kilogram body in terms of actual weight as I have done in Table 26. If we do that, we might almost tend to shrug away these minor components of the body. To avoid doing that, we must remember how tiny atoms are and how enormous even small weights (by normal standards) become when judged by atomic standards. Therefore, in Table 26, I also give the body content of each of these three long-lived radioactive nuclides in terms of numbers of atoms.

TABLE 26 — LONG-LIVED RADIOACTIVE NUCLIDES IN THE BODY

Nuclide	In 75-kilogram body	
	Weight (milligrams)	Number of atoms
Calcium-48	2,100	26,000,000,000,000,000,000,000
Potassium-40	37.5	560,000,000,000,000,000,000
Vanadium-50	0.015	200,000,000,000,000,000

See how the viewpoint changes in terms of numbers of atoms. If vanadium-50, the least of the three, were evenly divided among the roughly fifty trillion cells that make up the human being there would be 4,000 atoms of this very minor component of a very minor trace element *in each cell.*

The question is, though, whether the radioactivity of these nuclides affects the body.

To begin with, you might think that the radioactivity must affect the body, and perhaps very dangerously. Consider—

When a speeding particle or a gamma ray from a radioactive breakdown taking place outside the body happens to hit the body and streak through it, it will, most likely collide with a water molecule and break it up, leaving a very chemically active "free radical" behind. The free radical reacts with something else and may, indeed, initiate a chain of reactions until the energy of the initial particle or gamma ray is distributed and diluted and all is normal again. It is like the surface of a pond momentarily disturbed by the dropping of a pebble.

In general, such an event doesn't do serious damage, and, in fact, many billions of such events may take place, with the body absorbing and readjusting and retaining its balance. Every once in a while, however, the particle or gamma ray, or one of the energetic free radicals it produces, may collide with the DNA molecule of a gene. The gene will be altered, and to the extent that it controls some particular reaction in the cell, the chemistry of the cell will be altered.

The alteration may be harmless, but (just possibly) it could lead to the kind of damage of the growth-control mechanism that will make the cell cancerous in nature, or (just possibly) the alteration of the cellular chemistry may be extreme enough to kill the cell.

The striking of a DNA molecule is a low-probability event. Even lower in probability is the striking of a DNA molecule in the genes of those cells producing eggs or sperm. In that case, a mutation would be produced which affects not only a particular cell among the trillions in the body, but one which could find its way into a fertilized ovum and affect *all* the cells of a new organism.

Despite the utter unlikelihood of such an over-all mutation produced in one particular organism by one particular radiation event, if we consider all the organisms there are and have been and all the radiation events there are and have been, it becomes certain that such radiation-induced mutations will and must take place. Such radiation events must be a significant factor in the driving force behind evolution.

But when radiation originates outside the body, there is always the chance that it may miss a particular organism or, indeed, miss all organisms and expend itself on soil, water, or air. One can even imagine an organism accidentally or (if intelligent) deliberately shielding itself so as to be safe from most radiation.

Where radiation originates inside the body, however, how can one escape? Now we are dealing with nuclides that are, so to speak, firing at close range and cannot altogether miss. Every radioactive nuclide that is part of the body may break down at any moment and produce particles or photons that must plough through the molecules of the organism of which it is part.

Doesn't that seem dangerous? —And yet there are mitigating circumstances.

Each of the three long-lived radioactive nuclides that are to be found in our bodies emits a beta particle when it undergoes a radioactive breakdown. This, while not exactly salubrious, does less harm than an alpha particle would.

The next point of relief is that the longer the half-life, the fewer the radioactive breakdowns in any given time, and for calcium-48, the most common of the three, the half-life is the longest—20,000,-000,000,000,000 years or 20,000,000 eons, where an eon is equal to a billion years.

This means that, on the average, out of all the 26 sextillion carbon-48 atoms in the human body, there will be only 0.03 breakdowns per second; or, if you prefer, one every 33 seconds. This is a very small number as such things go and the damage we can expect calcium-48 to do is quite minimal. Almost everything we encounter in the world will supply us with more radioactivity than our body's own calcium-48 will.

The half-lives of potassium-40 and vanadium-50 are shorter than that of carbon-48, but the quantities are lower. The results for all three are presented in Table 27.

You can see at once that we can forget all about vanadium-50 and calcium-48. In comparison to potassium-40, the beta particles produced by the other two long-lived radioactive nuclides are excessively few. As far as we can see by the figures in Table 27, human radioactivity is produced just about entirely by potassium-40.

The activity of potassium-40 is nothing to get frightened over. Organisms have been living with this internal radioactivity throughout the history of life on this planet and the fact that life still flourishes is proof that we have successfully lived with it.

In fact, we might argue that potassium-40 is a driving factor in evolution and therefore very important to us in a beneficial way. Yes, but remember that there can be too much of a good thing. Evolution progresses by way of mutations, most of which are detrimental. The

TABLE 27 — RADIOACTIVE BREAKDOWNS IN THE BODY

Nuclide	Breakdowns per second
Vanadium-50	0.000007
Calcium-48	0.03
Potassium-40	8,500

species progresses at the cost of death to many individuals who are tried and found wanting. And if the mutation rate increases to too high a level, so many individuals are tried and found wanting that the species as a whole dies out. Remember that.

In any case, though, we don't have to worry about having too much of potassium-40. In fact, we might even argue that potassium-40 breakdowns are, after all, insufficient to serve as the driving force of evolution, that there is evidence that potassium-40 and evolution are not closely connected.

Potassium-40 has been constantly breaking down and decreasing in quantity ever since the Earth has settled down as a solid body. When life was in its very early stages, three billion years ago or so, the quantity of potassium-40 existing on Earth (and therefore in the average cell) was some four times as great as it is today. And yet I am not aware that anyone has ever maintained that the rate of evolution has been slowing down with time as the concentration of potassium-40 has declined.

Can it be that the close-range firing of potassium-40 is unimportant, compared to solar and cosmic radiation and to chemicals in the environment, as a producer of mutations and a driving force of evolution? Or is there something within the body that we have not yet considered?

What about short-lived radioactive nuclides? For the most part, these are so short-lived that the amount that builds up in the environment and can find its way into the body is insignificant by any standards; even on the atomic scale.

There is only one exception.

Cosmic-ray particles colliding with molecules in the atmosphere manage to knock neutrons out of atomic nuclei and these neutrons in turn collide with the nuclei of nitrogen-14 (the most common nuclide in the atmosphere) and form carbon-14.

The carbon-14 is radioactive and eventually breaks down, but it is continually being formed so some is always in the atmosphere. What is more, it was found to have a half-life of 5,730 years. This is short enough compared to the eons-long half-lives of the long-lived radioactive nuclides, but it is surprisingly long compared to the radioactive nuclides of the other light elements. The longest half-life for any radioactive oxygen or nitrogen nuclide, for instance, is 10 minutes.

The surprisingly long half-life of carbon-14 allows quite a bit to be built up in the atmosphere. Carbon in the atmosphere makes up part of the carbon dioxide molecule and 0.033 per cent of the atmosphere is carbon dioxide. Out of every 800 billion carbon dioxide molecules, *one* contains a carbon-14 nuclide instead of either of the stable carbon nuclides, carbon-12 or carbon-13.

It doesn't sound like much—1 out of 800,000,000,000 of a substance that makes up only 0.033 per cent of the atmosphere—but consider! The atmosphere as a whole has a mass of 5,100,000,000,000,000 tonnes,* so the carbon dioxide content of the atmosphere comes to 1,700,000,000,000 tonnes. Since the carbon atom makes up 0.273 of the total mass of the carbon dioxide molecule, the total mass of carbon in the atmosphere is 620,000,000,000 tonnes. If 1 out of every 800 billion carbon atoms is carbon-14, there is, in the atmosphere, some 0.78 tonnes of carbon-14, or 780 kilograms (1,700 pounds, if you insist).

A world-wide natural supply of 780 kilograms of carbon-14 still sounds pretty negligible, but 780 kilograms is equivalent to 33,000 trillion trillion atoms, and that's quite a few.

Plants incorporate carbon dioxide into their own structures and they don't discriminate against carbon-14. Animals eat plants and incorporate carbon-containing plant substances into their own tissues and don't discriminate against carbon-14. The result is that throughout actively living tissue there is the same proportion of carbon-14 (1 out of 800 billion) as in the atmosphere. To be sure,

---

* A tonne is equal to 1,000 kilograms, or about 1.1 ordinary American tons. I am more and more inclined to use the metric system exclusively in these essays.

this carbon-14 is continually breaking down, but as long as an organism is alive and metabolizing, new carbon-14 continually enters and a constant level is maintained.

Since 18 per cent of the human body is carbon, a 75-kilogram individual contains 13.5 kilograms of carbon and of this 0.000000017 grams is carbon-14. Put into terms of atoms, the 75-kilogram body contains 730,000,000,000,000 carbon-14 atoms.

This is only a little over a millionth the number of potassium-40 atoms in the body and if we suspect that potassium-40 breakdowns may not be significant as a driving force behind mutations and evolution, why should we worry about carbon-14?

Of course, carbon-14 has a half-life of less than six thousand years, so it breaks down at a much greater rate than the long-lived potassium-40 atoms. The relatively small number of carbon-14 in the body produces about 2,800 breakdowns per second, a surprisingly large quantity—but that still leaves carbon-14 only second best. Carbon-14 liberates only one third as many beta particles in the body as potassium-40 does. What's more, the beta particles produced by carbon-14 are only about one tenth as energetic as those produced by potassium-40. On the whole, then, you might expect potassium-40 to have the potential of doing some thirty times as much gene-twisting as carbon-14 can.

Yet carbon-14, with its considerably fewer and considerably weaker beta particles, can offer something far beyond anything possible for potassium-40. Potassium-40 is scattered through the body but none of it can actually form part of a gene. Carbon-14 is also scattered through the body but *some of it is inevitably found within a gene!* It fires at closest range!

Even though potassium-40 produces particles from its vantage point within the body, those particles may miss the genes and usually do. When carbon-14 is actually part of a gene, however, its breakdown makes a change in the structure of the gene *inevitable.*

You see, when a carbon-14 atom produces a beta particle, the atom itself recoils and does so energetically enough to break one or more of the bonds that holds it to its neighbor atoms. Nor can it possibly reconstitute those bonds after breakdown, for with the loss of the beta particle, the carbon-14 is converted into nitrogen-14 and this cannot take the place of the carbon-14.

The weight of carbon in the genes of a 75-kilogram body is about

250 grams, or 1.85 per cent of the carbon in the body. It follows that 1.85 per cent of the carbon-14 of the body is in the genes and that 1.85 per cent of the carbon-14 breakdowns in the body take place in the genes.

This means that every single second about fifty carbon-14 particles within your genes are breaking down and, therefore, fifty genes somewhere within your body are undergoing mutations of one sort or another.

In one way, this isn't much. If every such mutation took place in a different cell, then even after a lifetime of seventy years, only 2 per cent of your cells will have been affected. And yet—

How much of this ceaseless drizzle of mutations manages to put an occasional key cell out of action or produce an occasional biochemical malformation that has distant but important consequences? Does this necessary mutation rate contribute to the aging process, for instance, and, of course, to evolution?

Remember, too, that while the quantity of potassium-40 on Earth has been steadily dwindling and will continue to do so, carbon-14 remains more or less constant. Its concentration depends on such things as the cosmic-ray flux and the strength of the Earth's magnetic field (which serves to deflect cosmic-ray particles) and these go up and down erratically.

And even a small rise in carbon-14 levels might result in an important rise in mutation rate—to the point where many species might mutate themselves to extinction. With that in mind, consider that the radiation produced by nuclear bombs exploded in the atmosphere can produce carbon-14 in the same way as cosmic-ray particles can turn the trick.

How much carbon-14 was produced by the many nuclear explosions in the atmosphere over the last thirty years, I wonder, and how much of that got into living tissue?

Back in 1958, when atmospheric testing of nuclear bombs still went on wholesale, Linus Pauling (my favorite chemist) published a paper in Science (November 14, 1958) which went into the dangers of carbon-14 in a careful and systematic way. I'm sure this played its part in the eventual agreement on the part of the three chief nuclear powers to suspend atmospheric testing.

And yet the fact is that the very first mention of the role of carbon-14 in genetics, as far as I know, came three years earlier in my own article "The Radioactivity of the Human Body," which was published in *The Journal of Chemical Education* (Volume 32, pages 84–85, February 1955).

My own casual reference to this matter in my early article *in no way* takes precedence over Linus Pauling's detailed and cogent article in *Science*. Still, I have a letter from Professor Pauling, dated 11 February 1959, which refers in most kindly fashion to my article and—well, I just thought I'd mention it.

E—ENERGY

15. The Double-Ended Candle

The day after the President of the United States introduced the
nation to the "energy crisis" in the fall of 1973, that excellent writer
Barry N. Malzberg sent me a letter in which he said, "Nixon's
speech last night sounded like a brief extracted from one of your
articles five to seven years ago and I want you to know that I
thought of you all during it."

As it happens, I have been concerned with the limits of growth
for many years, but did not usually emphasize the energy aspect of
it. I spoke, instead, mainly of population growth, for that has
within it, as inevitable consequences, all the deadly crises that now
face us—not only a fuel shortage, but a general materials shortage,
including food. It brings along with it the prospect of a shattered
ecology, a poisoned biosphere, a raped planet, and a psychotic
humanity.

The first occasion on which I actually converted my fears into a
written essay was no less than seventeen years ago. I then wrote
"Fecundity Limited," which appeared in the January 1958 issue of
Venture Science Fiction and was afterward included in my collection *Is Anyone There?* (Doubleday, 1967).

Robert P. Mills, the editor of *Venture*, introduced my article as
follows: "With the death rate declining and sex here to stay, mankind is increasing his numbers at what may be an insane rate. Dr.
Asimov here examines some of the perils and problems we may
soon face."

In this article, I went through the routine (which I have repeated
on several occasions since) of calculating the consequences of a
continuing population increase at the current rate and demonstrating
how soon—how dreadfully *soon*—it would surpass any reasonable
limit that we can conceivably tolerate.

But that article was only 1,300 words long and I had no room to mention anything but population. When I began writing my articles for *F & SF*, I had three times the space at my disposal and eventually I decided to rewrite the earlier article at greater length. The result was "The Power of Progression," which appeared in the May 1969 *F & SF* and which was included in my collection *The Stars in Their Courses* (Doubleday, 1971).

I began that article by describing my own comfortable, middle-class, stodgy, compulsive-writer life and said: "What a pity, then, that it is all illusion and that I cannot blind myself to the truth. My island of comfort is but a quiet bubble in a torrent that is heaving its way downhill to utter catastrophe. I see nothing to stand in its way and can only watch in helpless horror."

Here again I concentrated on population—the fundamental problem out of which all else rises. However, in this article, which was 4,200 words long, I had space at least to mention some of the consequences. I said:

"What about resources? Already, with a population of 3.5 billion and the present level of technology, we are eroding our soil, spreading our minerals thin, destroying our forests, and consuming irreplaceable coal and oil at a fearful rate."

I went into still more detail in an article I wrote entitled "The End," which appeared in the January 1971 issue of *Penthouse* and which was included in my collection *Today and Tomorrow and . . .* (Doubleday, 1973). Here is what I said there:

"There is the matter of energy, for instance. Mankind has been using energy at a greater and greater rate throughout his existence . . .

"At the present moment, the total rate of energy utilization by mankind is doubling every fifteen years, and we might reasonably ask how long that can continue.

"Mankind is currently using energy, it is estimated, at the rate of 20,000,000,000,000,000,000,000 (twenty billion billion) calories per year. To avoid dealing with too many zeros, we can define this quantity as one 'annual energy unit' and abbreviate that as AEU. In other words, we will say that mankind is using energy, now, at the rate of 1 AEU a year. Allowing a doubling every fifteen years . . . you can calculate the rate of energy utilization in any given year and the total utilization up to that year.

"Right now, the major portion of our energy comes from the burning of fossil fuels (coal, oil, and gas), which have been gradually formed over hundreds of millions of years. There is a fixed quantity of these and they cannot be re-formed in any reasonable time.

"The total quantity of fossil fuels thought to be stored in the Earth's crust will liberate about 7,500 AEU when burned. Not all that quantity of fuel can be dug or drilled out of the Earth. Some of it is so deep or so widely dispersed that more energy must be expended to get it than would be obtained from it. We might estimate the energy of the recoverable fossil fuels to be about 1,000 AEU.

"If the 1,000 AEU of fossil fuels is all we will have as an energy source, then, at the present increase of energy utilization, we will have used it up completely in 135 years."

But, of course, the discussion in "The End" took up the matter of fossil fuels in general. What about oil specifically? In a small book I wrote in July 1973 for the Atomic Energy Commission (it is not yet published) I discussed oil, or petroleum, in particular and said:

". . . the growing dependence on superconvenient petroleum comes at a price. There is a considerably smaller supply of petroleum on Earth than there is of coal. At the present rate of use, the world's supply of petroleum may be scraping bottom in half a century at most.

"There is another complication in the fact that petroleum is not nearly as evenly distributed as coal is, and most of the industrial nations have become petroleum importers and are economically dependent on the producing nations. The United States, for instance, has 10 per cent of the total petroleum reserves of the world in its own territory and has been a major producer for decades. Its enormous and steadily growing consumption of petroleum, however, has outstripped its production capacity and made it an importer . . .

"Nearly three fifths of all known petroleum reserves on Earth are to be found in the territory of the various Middle Eastern countries. Kuwait, for instance, which is a small nation at the head of the Persian Gulf, with an area only three fourths that of Massachusetts and a population of about half a million, has petroleum reserves twice as great as those of the entire United States."

So for seventeen years, then, I have been writing—and lecturing —on the dangers of carelessly and continually expanding our numbers and our rate of energy use. It meant that I was lumped in with the category of the "doomcriers" and was shrugged aside by many as someone without faith in the ability of mankind to solve its problems.

That didn't bother me. I was quite certain that I was right and that I had therefore to continue to sound the warning, no matter what.

And then, all of a sudden, in November 1973, the "energy crisis" was upon us. Out of nowhere, out of a blue sky it seemed, we were suddenly facing disaster.

How did all this happen so quickly? From where did it come? To be sure, in mid-October the fourth war between Israel and the Arab nations broke out and the United States supported Israel, so the Arabs declared a boycott of oil shipments to the United States. But that meant a total cut in our energy supply of only 6 per cent. Was that enough of a reason to go into so deep a tailspin? Or is there something more?

And if there is something more, can it be that no one knew? Can it be that the energy crunch caught the leaders of this nation by surprise?

Nonsense! After all, *I* knew, and I didn't invent any figures. I didn't get them out of some arcane research methods of my own. I picked them up out of the public prints, out of readily available books, out of the essays I found in magazines and newspapers. If I could do it, anyone could do it.

The only conclusion I can come to is that many people in authority in both government and industry knew this was coming, but they didn't do anything and they didn't say anything. Why?

I have a theory about that if you're willing to listen. Just a theory, you understand, which, I am sure, is an oversimplification, but—

In the course of the first half of the twentieth century, the United States made the shift from coal to oil. In 1900 the energy derived from burning petroleum in the United States was only 4 per cent of that derived from burning coal. By the time World War II was over, we were getting more of our energy from oil and natural gas than from coal, with the balance shifting farther in favor of oil and gas each year.

Oil, being liquid, is much more convenient to mine, transport, and use, than coal is; much more. The switch to oil in the United States meant that energy was much more easily available. Electricity poured out of the nation's generators in an endlessly increasing stream and we began to live in a world in which all the controls were at our fingertips, so to speak.

And why not? The real size of the pools of oil underlying the Middle East only became known after World War II and we all received the impression of a hitherto-unknown ocean of oil ready for the taking. For a few heady years we used all we wanted, more and more and more, and it seemed to us that the supply was so great that we could postpone thoughts of a possible end to the some indefinite future. Let our grandchildren worry—and suffer.

The average American, I am sure, didn't think of an end at all. To him, electricity was just something you got out of a plug in the wall. But what about the leaders of our nation and of our economy? Did they also think that the glorious years would last forever?

How is it conceivable they could? They *had* to know; but for some reason they had to keep quiet.

What could it be? Was there anything else relevant that took place immediately after World War II in addition to the conversion of the nation to an economy based on endless oil-burning? There was the Cold War. How about that?

From 1945 onward, the United States was in stern competition with Soviet Union in a battle that did not make use of bombs and guns and planes and tanks. It was a battle just the same, with the whole world as the arena and the whole world as the prize. And since the weapons used were not of the military variety, they had to be political, economic, and psychological.

To those of us who, like myself, believe in the vision of America as presented by such men as Jefferson and Lincoln, it would seem that the great weapon of the United States was its freedom, its civil liberty, its absence of repression. Who, comparing the open society of the United States with the tightly controlled centralized socialism of the Soviet bloc could possibly choose, voluntarily, the latter?

It was not the weapon of Liberty, however, that the United States chose to use. Unfortunately, most people in the United States in the aftermath of World War II seemed highly suspicious of those among us who spoke too earnestly of Liberty. The mood was eventually seized on by Senator Joseph P. McCarthy and was carried by him to a peak that brought disaster to him and to the nation. We can therefore call it "McCarthyism."

Thanks to McCarthyism, it became possible, for instance, to circulate the Bill of Rights, word for word, in the form of a petition, and find that almost no one would sign it. Some thought that the words of the Bill of Rights represented dangerous communistic propaganda and others thought that signing any petition at all was a communistic act.

So we threw away the weapon of Liberty and took up, instead, one which we can call Affluence.

We were the richest nation on Earth, the one with the most advanced technology. We could supply loans and credits and did, and we knew the Soviet Union could not match us in that. We could supply technological knowhow and did, and we could not be matched in that either. What's more, we supplied this all to our friends and allies and denied it to our enemies and competitors, and made our doing so perfectly clear for it was the *selectivity* of our aid that was our weapon.*

Our weapon of Affluence worked at first, for the Soviet Union could not, in actual fact, match us. By means of the Marshall Plan and through foreign aid of all sorts, we rebuilt Western Europe and Japan, and we made the governments (if not always the people) of many underdeveloped nations comfortable.

So all of them became firm allies of ours, and all through the 1950s the United Nations delivered enormous majorities in favor of those positions supported by the United States. In fact, the Soviet Union had the votes only of itself and of those few nations it dominated and could only save itself from utter diplomatic defeat by the continual use of the veto in the Security Council.

But there is a difference between the weapons of Liberty and of Affluence.

Liberty is an absolute. It is not just the fact that you have *more* Liberty that counts; Liberty itself is enough. If the Soviet Union

---

* It was this use of our technology resource as an instrument of foreign policy that makes it impossible for us to object today to the Arab use of their oil resource as an instrument of their foreign policy. The Arabs are perfectly frank in saying that they are merely following the example we set them.

tried to counter such a weapon by developing Liberty of its own, then we would have won the Cold War, for the whole purpose of the Cold War was to see to it that the world would follow our system of government rather than that of the Soviet Union.

Affluence, on the other hand, is only relative. It was not important that we were affluent; only that we were more affluent than anyone else, and particularly that we were more affluent than the Soviet Union. If the Soviet Union gained our Affluence without adopting our Liberty than we would have lost the Cold War by losing our weapon.

It was that, I think, that created the absolute panic in the United States when the Soviet Union lofted Sputnik I into orbit on October 4, 1957. In itself, the feat was nothing to be alarmed about, but the world could not be allowed to think that the Soviet Union had outclassed American technology and that it was en route to an Affluence like ours or greater. Let the world think that and we would have lost the Cold War.

As a result, everything went into high gear and we grew determined to leapfrog the Soviet Union and reach the Moon first. Of course we did, and the Cold War was not lost—at least, not because of events in space.

But observe— If the United States depended on superiority in Affluence to win the leadership of the world, it could scarcely limit the rate at which energy was used by its citizens. On the contrary, it had to encourage it, and Americans generally leaped at the chance.

Partly, the drive for greater energy use was a natural consequence of our history. The United States grew up in an environment of an endless frontier, endless development, endless growth. It was hard to realize that there had to come an end to endlessness and that it had, in fact, come.

Add to this the natural consequence of an unregulated economy (Liberty has its price). The companies that sold oil and electricity to the nation naturally plunged for the short-term profits involved in encouraging the ever-greater use of energy, regardless of the long-term difficulties that entailed. Ditto, ditto, the automobile companies.

But the public might have been educated and the various companies brought to heel were it not for the fact that Affluence was so important a part of our foreign policy.

We deliberately built up West Berlin, for instance, as a show-piece of wealth in order that it might contrast with the stark dilap-idation of East Berlin, knowing that the Soviet Union could not conveniently match us, dollar for dollar. And it worked, for East Germany had to build a wall around West Berlin in the end, to hide the temptation and lure of wealth from its citizens and to close an easy route of escape for those to whose natural desire for liberty was added that temptation and lure.

Could we, then, spoil our successes by asking our own citizens to practice thrift and caution in accelerating the growth of our Affluence? In fact, so closely was our economic organization tied to endless growth that any halt in that growth, even a slowing down, would produce a recession.

So on we went to an extravagance of jet planes, and color television, and self-defrosting refrigerators, and self-cleaning ovens, and automobiles that powered every moving part, and buildings that were heated to 80° in winter and cooled to 60° in summer—and always, and in every case, at the cost of a further acceleration in energy use.

We developed a higher and higher energy-rate economy and way of life. Between the end of World War II and the beginning of the 1970s we lived through an amazing quarter century. The population of the United States increased by one and a half times, while the rate of energy use went up three times. The 210,000,000 Americans today use, twice as much energy per capita as did the 140,000,000 Americans of yesterday.

In 1920 Edna St. Vincent Millay, in a well-known verse, spoke of candles burning at both ends which wouldn't last the night. She did admit, though, that they gave a lovely light.

Well, from 1945 to 1973 the United States burned its candle at both ends and it *did* give a lovely light. But surely we could have foreseen that it would not last the night.

In that period of time during which our total rate of energy use tripled, we went from being an oil-exporting nation to an oil-im-porting nation. And we had to import oil in competition with Japan and the West European nations, all of whom we had helped to Affluence of their own.

It became harder and harder to maintain a favorable balance of trade and the dollar weakened, particularly with respect to the West German mark and the Japanese yen. (The thing was we barred the defeated Germans and Japanese from wasting their money on supporting a military machine. We were determined to waste enough of our own money to support enough of a military machine to defend them as well as ourselves. Then, too, we developed the military machine itself into a higher and higher energy-using device so that it became a more and more affluent object even in peacetime.)

Yet, as the hole we were digging for ourselves became more and more apparent throughout the 1960s, we could still do nothing to conserve our position by controlling our rate of energy use. Our foreign policy and the public devotion to the energy-rich way of life made it impossible for any political leader to advocate energy control.

Failing such energy control, it would have made good sense either to keep the Arab nations in subjection or to cultivate them as friends. We did neither. Why not? Partly, of course, it was because we were committed to supporting Israel, but this was by no means the full explanation. The fact is that during the crucial decade of the 1960s, the United States had its attention diverted by our greatest folly since the Civil War.

The period of McCarthyism had convinced so many American people that we had "lost China" through treason in high places, that it became politically impossible for any American President to "lose" any other region, even when it would have benefited us to abandon areas of minimal importance in order to conserve our strength for utilization at crucial points.

After World War II the United States adopted the principle of no retreat, anywhere; something like the stand which Hitler had taken in World War II; and with the results you would expect of anything so inflexible. We wasted our strength on useless struggles in positions chosen by our adversaries. We fought in Korea to a draw at worst; and then we fought in Vietnam to a draw at best.

The Vietnam War occupied us for twelve years uselessly. North Vietnam and the Viet Cong are stronger today than they were when we started.

What's more, the wars we fought in the 1950s and 1960s were high-energy wars as befitted our new American way of life. American soldiers never walked, they helicoptered; they ate and lived in as much luxury as we could manage. Naturally, it is hard to begrudge them this, but consider—

During the Korean War we used to complain about the Chinese "hordes" who attacked our men. When we were pushed back, we insisted, it was only by overwhelming numbers.

But it is not only numbers that can overwhelm. We attacked our Asian enemies with "hordes" of energy and when the enemy fell back they might justly claim to do so before overwhelming calories. —And the fact is that in the last analysis they could afford the men better than we could afford the calories (short of nuclear weapons which, out of regard for world opinion, we dared not use) and we had to pull out short of victory.

So while our Presidents dared not "lose" South Korea or South Vietnam, and none dared "be the first President of the United States to lose a war," we lost the Middle East and the Cold War. How could we take strong stands against Soviet domination of the Middle East when, throughout the 1960s, our eyes were fixed entirely on Vietnam—despite the fact that nowhere in the world could one find an area of comparable size that is farther from us and less important to our welfare? And how can we take strong action now that we are out of Vietnam, when the memory of that traumatic war prevents us taking such strong action anywhere.

No, the price of Vietnam for the 1970s was that the China we so bitterly opposed had become strong enough to enter the United Nations to the triumphant applause of delegates from all over the world, including many who had profited from our proffered Affluence. And furthermore, the Soviet Union had become strong enough militarily, economically, and diplomatically to force us into an attitude of détente rather than opposition.

Yet with all this, we *still* dared not puncture our energy-wasting economy.

Our cities were beginning to experience brownouts at every heat wave and cold wave, but the oil companies, intent on their short-term profits, were content to blame this on the conservationists, those few wicked people who wanted clean air. The fact that the oil companies had encouraged an energy-wasting economy for profits and that the government had done so for the sake of a foreign policy that did not work was never mentioned.

Even our high-energy military machine was beginning to suffer. It was so incredibly wasteful of energy that we were having difficulty keeping it up to snuff even under peacetime conditions. Nuclear wars could not be fought because that was suicide; and the fact is that non-nuclear wars cannot be fought either, because they are too expensive. The Vietnam War had strained us dangerously and we could scarcely find a smaller enemy than North Vietnam.

Yet *still*, through the summer of 1973, we dared not do anything about our arrow-straight flight to catastrophe.

Then, in October 1973, came the Arab oil boycott and a 6 per cent reduction of our energy supply. It was enough. At last, at last, at last the government could say "crisis" with no blame attached to itself.

Suddenly—overnight—the United States bloomed into a gasoline shortage, a fuel oil shortage, a paper shortage, a plastics shortage, an adhesives shortage—a shortage, in short, of everything but shortages.

The only way of understanding this is to remember that the energy crisis was there all along. The Arab boycott did not cause it; it only made it possible to stop faking its absence.

The Arabs did us a favor, then. They have made it possible for us to tackle the problem now, when perhaps it is not too late. The danger we really face is that the Arabs will relent and offer us oil and that we will be fools enough to take it and go back to burning our candle at both ends—and prepare for ourselves a far worse catastrophe ten years hence.

But assuming we have learned our lesson, however unwillingly, and that we now know that energy is a resource to be used carefully —what do we do now?

I'll consider that situation in the next chapter.

16. As Easy As Two Plus Three

I used to refer to myself, years ago, with as much modesty as I could manage (I never could manage much), as a "minor celebrity." By that I meant that perhaps one person out of a thousand thought I was a great and famous man, and the other nine hundred and ninety-nine had never heard of me.

That was just the right proportion, too. I manage to meet that one-out-of-a-thousand much oftener than chance alone would dictate (by going to science fiction conventions, for instance) and I can then bask in his (or, preferably, her) adulation. Aside from that, I retain my privacy and anonymity through much of my life, thanks to the other nine hundred and ninety-nine. This means that I don't have to worry if my hair is mussed up, as it usually is. I know that no one is going to nudge his neighbor and say, "Look at that great and famous man walking around with his hair mussed up."

But now this business of being a minor celebrity is getting out of hand. When I received in the mail a copy of the syndicated newspaper magazine section "Family Weekly" for 30 December 1973, I found there a column entitled "What in the World!", which listed, among other things, those people who were to have birthdays in the coming week.

It turned out that Sandy Koufax was thirty-eight and Bert Parks fifty-nine on Sunday, December 30. Then, too, Barry Goldwater was sixty-five, Xavier Cugat seventy-four, and Dana Andrews sixty-two on Tuesday, January 1. And then, to my mingled surprise and horror, I found listed for Wednesday, January 2, none other than your friend and mine Isaac Asimov. Worse yet, my age was given. Still worse, it was given correctly, as ——. Inevitably, I later got a rash of fan letters wishing me a happy birthday. (See also the beginning of Chapter 8.)

With things like that in the public prints, how am I going to be convincing when I tell all those beautiful young ladies at the science fiction conventions that I'm just a little past thirty—

It is a relief to turn to less serious things, then, like the energy shortage which is threatening civilization with destruction.

Suppose we make up our minds that cheap oil is gone forever and that the oil wells will be going dry by the end of the century. What then?

We can switch back to coal, of which a several-centuries supply remains in the ground. Digging it up at a considerably increased rate will, however, involve a great deal of damage to the environment; and will require the hiring of a great many men for the hazardous and unpopular task of wrenching the coal out of the ground. It will make necessary the shipping of coal in unprecedented quantities over a railway network we have carelessly allowed to go to ruin; or else we must develop large-scale (and expensive) methods for the conversion of coal into liquid and gaseous fuel on-site. The same difficulties can be listed for the squeezing of oil out of shale.

Still, this may serve as a stopgap while science and technology learn to make greater use of such energy sources as wind, tides, running water, geothermal heat, and sunlight.

My own feeling, however, is that the best and most versatile energy source of the future is nuclear in nature.

We have a kind of working nuclear energy source already, of course. For a generation, men have been splitting uranium nuclei ("nuclear fission") and have obtained useful energy as a result. At the present moment a "breeder reactor" capable of producing more fission fuel than it consumes has been put into action by the Soviet Union.

Other nations, including the United States, have breeder reactors in prospect and, by such means, the Earth's entire available supply of two heavy metals, uranium and thorium, can be used as a source of energy. Fission energy by way of breeder reactors could last mankind for something like a hundred thousand years.

That sounds pretty good. It's a lot and it's here—but there's a catch. Although nuclear fission doesn't produce the kind of air pollution that coal and petroleum do, the split fragments of the heavy atoms undergoing fission are radioactive.

This radioactivity is far more dangerous than an equivalent amount of ordinary chemical pollution can be, and it must not be allowed to enter the environment until the radioactivity dies away—which can take centuries. Any reliance on nuclear fission, then, will require the safe and permanent disposal of dangerously radioactive fission products, which will be produced at a steadily increasing rate.

Then, too, there is always the small chance that a fission reactor may go out of control and send radioactivity broadcast over a wide region. So far, truly serious accidents have not occurred, but it is impossible to guarantee that accidents will *never* happen, so that there is a considerable resistance on the part of some parts of the population to the further expansion of fission energy.

The solution? —Fusion energy.

It is the medium-sized atomic nuclei that have the least energy. Splitting large nuclei into those nearer the medium size (fission) is one way of getting energy. Another is to start at the other end of the list of atoms and combine very small nuclei into those that are larger and nearer the medium size.

The process of forcing small atomic nuclei to fuse into larger ones is "nuclear fusion." The best known example of this in nature is the process whereby the nuclei of hydrogen atoms (the smallest variety known) are made to undergo fusion in the core of stars, producing nuclei of helium atoms (the second smallest) and also producing the vast quantities of energy which stars, including our own Sun, pour out into space in all directions.

Nuclear fusion offers an even richer supply of energy than nuclear fission does. A kilogram of hydrogen undergoing fusion to helium will produce at least four times as much energy as a kilogram of uranium undergoing fission. Furthermore, hydrogen is a much more common substance than uranium so that the total fusion energy potentially available is enormously greater than the total fission energy available.

In the Universe as a whole, hydrogen is by far the most common substance. Nine tenths of all the atoms in the Universe, astronomers estimate, are hydrogen. It is not surprising, then, that it is hydrogen fusion that powers the stars. It's the only thing that can, in fact. Nothing else is common enough to serve as fuel on so grand a scale. In our own Sun, 600,000,000 tonnes of hydrogen are undergoing fusion into helium every second! It is this which has supplied the vast energy output of the Sun through every second of its existence as a star over the past five billion years and more. What's more, the Sun is so huge and its hydrogen content so vast that there is enough to keep it going for billions of additional years.

Hydrogen fusion however is much more difficult to get started than uranium fission is. For uranium nuclei, a mildly energetic neutron is enough to turn the trick. Hydrogen nuclei, in order to fuse, must slam into each other with enormous energies. This is equivalent to saying that the hydrogen must be heated to enormous temperatures. Then, too, there must be a great many hydrogen nuclei in the vicinity, so that the heated nuclei won't miss each other too often as they tear around. This is equivalent to saying that the hydrogen must also be present at enormous densities.

The two requirements, high temperature and high density, are very difficult to achieve together. As the temperature goes up, the hydrogen tends to expand so that its density tends to decrease dramatically. Long before the temperature has reached the level where we can have any hope of fusion at the original density, the hydrogen (under ordinary conditions) has spread out so thinly that there is no chance of fusion at any temperature.

This means that the hydrogen must be confined while it is being heated. The best way of confining it is by means of an enormous gravitational field. This is what happens on the Sun. The Sun's own powerful gravity compresses its innermost core to the point where the hydrogen is hundreds of times as dense there as it is on Earth and keeps it that dense even though the temperature at the Sun's center reaches some 15,000,000° C.

There is no chance, however, that we can duplicate this sort of "gravitational confinement" on Earth. A gravitational force like the Sun's, or anything like it, simply cannot be produced in the laboratory.

Furthermore, we can't deal with hydrogen at densities such as those which exist at the Sun's core. We have to deal with hydrogen at much smaller densities and make up for this by achieving temperatures even higher than those that exist in the Sun's core. But then, as we try to reach such enormous temperatures, what is to keep the hydrogen from expanding at once into the thinnest, most useless wisps?

Is there any way in which we can heat the hydrogen so quickly that before the hydrogen atoms have time to expand they are already at the temperature needed for fusion? The time it takes for hydrogen to expand is greater than zero because of the principle of inertia, so the process by which hydrogen atoms are kept in place simply because they lack the time to expand is said to be "inertial confinement."

In the 1950s the only way by which heat could be produced quickly enough to raise hydrogen to the necessary ignition temperature before it could expand was to make use of a uranium-fission bomb. When this was done, the fission igniter started the hydrogen-fusion reaction and produced a much vaster explosion. The result is the so-called "hydrogen bomb" or "H-bomb," which is more appropriately called a "fusion bomb."

The first fusion bomb was exploded in 1952, but though that demonstrated that nuclear fusion was possible on Earth, the demonstration was that of uncontrolled fusion and what we need in order to have energy in peace is *controlled* fusion.

Can we find some way to confine hydrogen while we heat it slowly—some system of confinement other than gravitational, some system we can handle in the laboratory? Or else, can we heat hydrogen very quickly to take advantage of inertial confinement by some means other than that of a fission bomb?

Gravitational confinement can be replaced by "magnetic confinement." A magnetic field is enormously stronger than a gravitational one and can be handled easily in the laboratory. One disadvantage is that whereas a gravitational field works on all matter, a magnetic one works only on electrically charged matter—but that is easily taken care of. Long before fusion temperatures are reached, all neutral atoms are broken up into electrically charged fragments ("plasma").

The electrically charged particles of plasma are influenced by a magnetic field. One of the proper kind and shape will cause the particles of a plasma to move in certain directions that will keep them within the bounds of the field. The particles of the plasma will then be confined by the insubstantial shape of the field. (Naturally, no material object can confine the superhot plasma—since the plasma will make contact with the confining matter, and either the plasma will cool down or the matter will vaporize.)

The difficulty with magnetic confinement is this. The magnetic fields set up to confine the plasma are unstable. After a very tiny fraction of a second, they can develop a leak or change their shape and, in either case, break up so that the confined plasma instantly expands and is gone.

For a quarter of a century nuclear scientists in both the United States and the Soviet Union have been trying to work out ways for producing a magnetic field of such a shape that it will stay stable long enough to allow a sample of plasma which is dense enough for fusion to become, also, hot enough for fusion.

Suppose, for instance, that an electric current is made to flow through a plasma which is within a glass cylinder. The current sets up a magnetic field that fits around the plasma like a series of circles. These circles tend to produce a "pinch effect," pinching the plasma inward away from any fatal contact with the glass. Within the magnetic field, the plasma can then be heated, but if the field twists or leaks, the plasma would flow out, touch the glass, cool down at once and you would have to start over.

Naturally, we wouldn't want the plasma to leak out the ends of the cylinder. One way of preventing that is to build up particularly strong magnetic fields at either end of the cylinder, fields that would repel particles reaching them and reflect them backward. Such an arrangement is a "magnetic mirror."

Another way is to insert the plasma into a doughnut-shaped cylinder so that the plasma will go round and round and there will be no ends out of which to escape. Unfortunately, a magnetic field so designed as to enclose a doughnut of plasma is particularly unstable and would not last more than the tiniest fraction of a second.

To increase the stability, the doughnut cylinder was first twisted into a figure 8. This made it possible to produce a stabler magnetic field. Such an instrument was called a "stellarator," from the Latin word for "star," since it was hoped to duplicate, within it, the fusion reactions that took place in stars.

Then it was found that a simple doughnut was preferable after all if *two* magnetic fields were set up, the second in such a way as to stabilize the first. Soviet physicists modified the manner in which the second field was set up and improved its efficiency in a type of device they called "Tokamak," an abbreviation of a complicated Russian phrase. After the first Soviet results in Tokamak devices were announced in 1968 American physicists promptly modified some of their stellarators, incorporating the Tokamak principle.

It looks as though the Tokamak may suffice to bring about fusion if we don't ask too much of it. Let's see how we can ask the least.

Although it is hydrogen that physicists are trying to fuse, there are three varieties of hydrogen atoms. One has, as its nucleus, a single proton and nothing more, so it is called "hydrogen-1." Another has a nucleus of two particles, a proton and a neutron, so it is called "hydrogen-2" (or deuterium). The third has a three-particle nucleus, one with a proton and two neutrons, so it is called "hydrogen-3" (or tritium).

All three varieties of hydrogen will fuse to helium atoms. However, hydrogen-2 will fuse more easily and at a lower temperature than hydrogen-1, while hydrogen-3 will do so still more easily and at a still lower temperature.

In that case, why not forget all about ordinary hydrogen, shift to hydrogen-3 at once, and use it as our fusion fuel? Too bad; there's a catch.

Hydrogen-3 is not a stable substance. It is radioactive and breaks down rapidly so that hardly any of it exists on Earth. If we want to use hydrogen-3, with its twelve-year half-life, we have to form it by energy-consuming nuclear reactions. We would have to form it, for instance, by bombarding the light metal lithium with neutrons from a fission reactor. It would mean using an expensive fuel, difficult to produce and difficult to handle.

What about hydrogen-2? This is a stable atom and exists on Earth in considerable quantity. To be sure, only one atom of hydrogen out of seven thousand is hydrogen-2 (virtually all the rest are hydrogen-1). Still, there is so much hydrogen on Earth that even one out of seven thousand isn't bad.

For instance, every liter of water contains about ten thousand billion billion atoms of hydrogen-2, and if all of this deuterium is made to undergo fusion it will deliver as much energy as 300 liters of gasoline. Since there are nearly three thousand billion billion liters of water on Earth, it is easy to see that there would be enough fusion energy to last mankind, at the present rate of energy-use, for many billions of years.

But though hydrogen-2 is easier to ignite than hydrogen-1, the ignition is not really easy. At reasonable densities, temperatures of

400,000,000° C would be required. (This is twenty-five times the temperature of the Sun's interior where the even-more intractable hydrogen-1 undergoes fusion, but the density at the Sun's interior is enormous.)

We can compromise by using a mixture of hydrogen-2 and hydrogen-3, going to the expense and difficulty of manufacturing hydrogen-3, but in lesser amounts than if that were our only fuel.

The ignition temperature for a half-and-half mixture of hydrogen-2 and hydrogen-3 is only 45,000,000° C. This is the lowest known ignition temperature for any fusion reaction that involves at least some stable nuclei. It looks as though 2+3 is as easy as we can get in fusion.

Since the helium atom which is produced in hydrogen fusion has four particles in the nucleus (helium-4), the fusion of hydrogen-2 and hydrogen-3 leaves one particle over: 2+3=4+1. That one particle is a neutron.

As it happens, hydrogen-3 is formed by bombarding lithium with neutrons, as I said earlier. Therefore, although we might prepare hydrogen-3, to begin with, by using neutrons from fissioning uranium, once we got the 2+3 fusion reaction going, it itself would supply the neutrons for the formation of more hydrogen-3.

So what we need now is a 2+3 mixture in a Tokamak which is dense enough and hot enough to ignite.

In 1957 the British physicist J. D. Lawson had calculated that in order for ignition to take place once the proper temperature was reached, that temperature had to be maintained for a certain period of time. This length of time serves to make sure that enough nuclei collide in their random movements to produce enough heat to keep the reaction going. Naturally, the more dense the plasma, the more collisions will take place in a given time and the shorter the time of confinement necessary for ignition.

For hydrogen at the density found under ordinary conditions on Earth, the ignition temperature need be maintained for only four millionths of a second. Even the best magnetic devices for confining plasma can't hold hydrogen at that density for even that length of time. The density must be decreased therefore and that means that the time during which the temperature must be maintained must be correspondingly increased.

Soviet and American physicists have pushed the combined density and temperature time to higher and higher levels but have not yet reached ignition point. As magnetic fields continue to be strengthened and to be more subtly designed, it seems certain, though, that within a few years (how many exactly it is difficult to predict) the fusion fire will catch.

And, ironically, after a quarter century of work on electromagnetic confinement as a substitute for gravitational confinement, it seems that a later competitor in the field will overtake the magnetic field and prove to be the answer. We are back to inertial confinement with something replacing the atomic bomb as a superfast raiser of temperature.

In 1960 the laser was invented, a device by means of which radiant energy, such as visible light, could be delivered in large quantities concentrated into a very fine focus. For the first time, men learned how to deliver energy with a rapidity similar to that of a fission bomb, but on a so much smaller scale that the delivery was safe. (The first lasers were feeble indeed, but in the years since their invention, they have been growing steadily more powerful and more versatile.)

Suppose a laser beam is concentrated on a pinhead-sized pellet made up of a mixture of frozen hydrogen-2 and hydrogen-3. (This would be an extremely cold pellet, of course, since these substances don't freeze till a temperature of $-259°$ C, or $14°$ K—only 14 degrees above absolute zero—is reached.)

Naturally, the laser beam would evaporate the pellet in an extraordinarily tiny fraction of a second. However, as the energy from the laser beam continues to pour into the evaporating gas, the plasma that results is heated to ultrahigh temperatures in less time than it takes for the individual nuclei to move away. It is a case of inertial confinement, like that produced by the fission bomb, but on a small and controllable scale. Since the atoms have no time to move away, an enormous pressure is placed on the interior of the pellet which compresses to ultrahigh densities and ignition can take place.

In 1968 the Soviets first used this laser heating of solid hydrogen to the point of detecting individual fusion reactions among the nuclei, but not enough to produce actual ignition. Since then, both the United States and the Soviet Union have vastly increased the money being invested on this approach and now something like $30 million a year is being spent by each nation.

Plans are under way to build better and more energetic lasers and to focus a number of them on the frozen pellet from different directions.

Frozen hydrogen-2 and hydrogen-3 have densities about a thousand times that of the ordinary gaseous form to begin with. Compression of the central region of the pellet will increase that density an additional ten thousand times. Under such conditions it would take only a trillionth of a second after reaching the necessary temperature to set off a flood of fusion energy. It additional pellets are dropped into the chamber one after another, the flood could be made continuous.

Naturally, a great deal of energy must be invested in bringing about ignition. Enormous laser beams, consuming vast quantities of electricity are required, to say nothing of the energy required to isolate hydrogen-2 from ocean water and to form hydrogen-3 by neutron bombardment of lithium—and then to freeze both hydrogen-2 and hydrogen-3 to $14°$ K.

This energy investment is needed only at first, however. Once the fusion reaction is ignited, it will itself produce all the energy needed to keep things going.

The science involved seems certain to work—and within a few years, perhaps. Then will come the engineering end of it—the actual setting up of practical devices that will take the heat of fusion and convert it into useful energy.

If we begin by producing what will certainly be only a "first-generation" fusion reactor, with a mixture of hydrogen-2 and hydrogen-3 ignited by laser action, most of the energy produced will be in the form of very energetic neutrons. These neutrons will move outward from the fusion reactor in all directions and will strike a shielding shell of liquid lithium.

The neutrons will react with the lithium to produce hydrogen-3, which can be isolated and fed back into the fusion reactor. The lithium is heated, in the process of absorbing and reacting with the neutrons, and is cooled off by circulation past water reservoirs. The water is itself heated in the process and is converted to steam, which can turn a turbine and produce electricity in the conventional manner.

The advantages are enormous.

For one thing, the energy is endless, for the fuel (hydrogen-2 and lithium) exists in sufficient supply on Earth to last mankind for an indefinite period. The energy is ubiquitous, too, since the fuel exists all over the Earth.

The energy is safe since the only dangerous products are neutrons and hydrogen-3, and both are consumed if the reaction works with complete efficiency. What is left over is helium, which is completely safe. Nor is the danger, in case of accidental leakage anywhere, nearly what it would be in the case of nuclear fission.

In "second-generation" fusion reactors the danger would even be less. Physicists speculate that the hydrogen-2/hydrogen-3 fusion reaction can serve merely as an igniter to further fusion. You would have a pellet made of boron-11 and hydrogen-1, with hydrogen-2 and hydrogen-3 at the core. Under laser activation, the core would ignite and create energies high enough to set up a boron-hydrogen fusion ($3,000,000,000°$ K would be required).

The boron, fusing with hydrogen-1, would split into three nuclei of helium-4 ($11+1=4+4+4$) and nothing else—no neutrons, no radioactive particles of any nature.

Finally, there can be no explosion. In nuclear fission, energy can only be produced if large quantities of uranium are present in the fissioning core. Fusion, however, works with tiny quantities of fuel. If anything at all goes wrong, it can only succeed in stopping the process at once. Fusion energy presents us with an automatic fail-safe sytem.

But watch out—

First, fusion produces heat that would not otherwise exist on Earth ("thermal pollution"), just as burning coal and oil do. This additional heat, over and above that which arrives from the Sun, cannot be radiated away by the Earth without a slight rise in the planet's over-all temperature. If the availability of fusion energy encourages us to use it in such wild amounts as to increase energy production to perhaps a thousand times the present rate, we may succeed in melting the polar icecaps over a relatively small stretch of decades. This will drown the heavily populated coastal areas of the continents under two hundred feet of ocean water—and that will make breathing difficult.

Then, too, if the availability of fusion energy encourages mankind to harbor false illusions of security and to continue to multiply its numbers endlessly, our rather delicately balanced technological civilization will collapse for reasons other than energy shortage.

Finally, even if we ignite a fusion reaction tomorrow, the engineering problems involved in building practical large-scale power stations are enormous, and I would guess it would take thirty years—by which time the world population would be (barring castastrophe) 7,000,000,000.

So we've got at least thirty years ahead of us in which we're going to have to live under energy-short, population-high conditions. Mankind must walk a thinning and rising tightrope to reach possible safety. The odds are against us, but let's try.

F—NUMBERS

17.   Skewered!

I don't write many mathematical articles in this series, and for a very good reason. I don't have a mathematical mind and I am not one of those who, by mere thought, finds himself illuminated by a mathematical concept.

I have, however, a nephew, Daniel Asimov by name, who *does* have a mathematical mind. He is the other Ph.D. in the family and he is now an Assistant Professor of Mathematics at the University of Minnesota.

Some years ago, when he was yet a student at M.I.T., Danny had occasion to write to Martin Gardner and point out a small error in Gardner's excellent "Mathematical Recreations" column in *Scientific American*. Gardner acknowledged the error and wrote me to tell me about it and to ask a natural question. "Am I correct in assuming," said he, "that Daniel Asimov is your son?"

Well! As everyone knows, who knows me, I am only a little past thirty right now and was only a little past thirty at the time, some years ago, when this was taking place. I therefore wrote a letter to Gardner and told him, with some stiffness: "I am not old enough, Martin, to have a son who is old enough to be going to M.I.T. Danny is the son of my younger brother."

Friends of mine who have heard me tell this story keep assuring me that my statement involves a logical contradiction, but, as I say, I do not have a mathematical mind, and I just don't see that.

And yet I must write another mathematical article now because over eleven years ago I wrote one* in which I mentioned Skewes' number as the largest finite number that ever showed up in a mathematical proof. Ever since then, people have been asking me to write

---

* See "T-Formation," reprinted in *Adding a Dimension* (Doubleday, 1964).

an article on Skewes' number. The first request came on September 3, 1963, almost immediately after the article appeared. On that date, Mr. R. P. Boas of Evanston, Illinois, wrote me a long and fascinating letter on Skewes' number, with the clear intention of helping me write such an article.

I resisted that, along with repeated nudges from others in the years that followed, until March 3, 1974, when, at Boskone 11 (a Boston science fiction convention at which I was guest of honor), I was cornered by a fan and had Skewes' number requested of me. So I gave in. Eleven years of chivvying is enough.† I am Skewered.

First, what is Skewes' number? Not the numerical expression, but the significance. Here's the story as I got it from Mr. Boas (though I will paraphrase it, and if I get anything wrong, it's my fault, not his).

It involves prime numbers, which are those numbers that cannot be divided evenly by any number other than themselves and one. The numbers 7 and 13 are examples.

There are an infinite number of prime numbers, but as one goes up the list of numbers, the fraction of these numbers that are prime decreases. There is a formula that tells you the number of primes to be found in the list of numbers up to a given number, but like everything else about prime numbers, the formula is not neat and definite. It only tells you approximately how many primes there will be up to some limiting number.

Up to the highest limit that has actually been tested, it turns out that the actual number of primes that exist is somewhat *less* than is predicted by the formula.

In 1914, however, the British mathematician John Edensor Littlewood demonstrated that if one lengthened out the string of numbers which one investigated for primes, one would find that up to some limits there would indeed be less than the formula predicted but that up to other limits there would be more than the formula predicted.

---

† I'll admit that I've been chivvied longer than that in some respects. For seventeen years I have been requested, with varying degrees of impatience, to write another Lije Baley novel; and for over twenty years to write another *Foundation* novel. So please don't anybody write letters that begin with "If eleven years of chivvying is enough, why don't you——." Because I'm doing all I can, that's why.

In fact, if one continued up the line of numbers forever, the actual total number of primes would switch from less than the formula prediction to more than the formula prediction to less than the formula prediction, and so on—and make the switch an infinite number of times. If that were *not* so, Littlewood demonstrated, there would be a contradiction in the mathematical structure and that, of course, cannot be allowed.

The only trouble is that as far as we have actually gone in the list of numbers, not even one shift has taken place. The number of primes is always less than the formula would indicate. Of course, mathematicians might just go higher and higher up the list of numbers to see what happens, but that isn't so easy. The higher one goes, the longer it takes to test numbers for primehood.

However, it might be possible to do some theoretical work and determine some number below which the first switch from less than the prediction to more than the prediction *must* take place. That will at least set a limit to the work required.

Littlewood set S. Skewes (pronounced in two syllables by the way, Skew'ease) the task of finding that number. Skewes found that number and it proved to be enormously large; larger than any other number that ever turned up in the course of a mathematical proof up to that time, and it is this number which is popularly known as "Skewes' number."

Mind you, the proof does not indicate that one must reach Skewes' number before the number of primes shifts from less than the prediction to more. The proof merely says that some time *before* that number is reached—perhaps long long before—the shift must have occurred.

A number as large as Skewes' number is difficult to write. Some shorthand device must be used and the device used is the excellent one of exponential notation.

Thus, $1,000 = 10 \times 10 \times 10$, so 1,000 can be written as 10^3 (ten to the third power) where the little 3 is called an "exponent." The little 3 signifies that 1,000 can be considered the product of three 10s, or that it can be written as a 1 followed by three 0s. In general, 10^x (ten to the xth power) is the product of x 10s and can be written as a 1 followed by x zeros.

Since 10,000,000,000 is written as a 1 followed by 10 zeros, it can be written exponentially as 10^{10} (ten to the tenth power). In the same way a 1 followed by ten billion zeros, something that would be impractical to write, can be easily expressed exponentially as $10^{10,000,000,000}$ (ten to the ten billionth power). But since ten billion is itself 10^{10}, $10^{10,000,000,000}$ can be written, even more briefly, as $10^{10^{10}}$.

Writing exponentials is always a strain when an article is being written for a non-specialized outlet. This is especially so when one is forced to place exponents on exponents. To avoid driving the Noble Printer crazy and to make the notation look prettier, I have invented a notation of my own. I make the exponent a figure of normal size and it is as though it is being held up by a lever, and its added weight when its size grows bends the lever down, Thus, instead of writing ten to the third power as 10^3, I will write it as $10\backslash3$.

In the same way, ten to the ten billionth power can be written as $10\backslash10,000,000,000$, or as $10\backslash10\backslash10$.

Using this "Asimovian exponential notation," Skewes' number becomes $10\backslash10\backslash10\backslash34$.

Now let's see what Skewes' number might be in ordinary non-exponential notation. To do that we must consider the components of the exponential notation from right to left. Starting at the right, we know what 34 is, we move leftward and consider $10\backslash34$. This is ten to the thirty-fourth power and can be written as a 1 followed by 34 zeros, thus: 10,000,000,000,000,000,000,000,000,000,-000,000 or, in words, ten decillion (American style). This means that Skewes' number can be written $10\backslash10\backslash10,000,000,000,000,000,-000,000,000,000,000,000$.

So far, so good, if a bit disconcertingly formidable. The next step is to move one place to the left and ask how we might write: $10\backslash10,000,000,000,000,000,000,000,000,000,000,000,000$. Easy. You just put down a 1 and then follow it by ten million billion billion billion (or ten decillion, if you prefer) zeros.

If you were to try to write such a number by beginning with a 1 and then writing ten decillion zeros, each the size of a hydrogen atom, you would require nearly exactly the entire surface of the Earth to write the number. Furthermore, if you wrote each zero in a trillionth of a second and kept it up at that rate without cessation, it would take a thousand trillion years to write the entire number.

Anyway, let's call this number the "Earth-number," because it takes the Earth as a blackboard to write it, and imagine that we can write it. Now we can write Skewes' number as 10\Earth-number, and this means we now know how to write Skewes' number in the usual fashion. We start with a 1 and then follow it with an Earth-number of zeros.

This is tremendously more than the ten decillion zeros it took merely to write the Earth-number. A number itself is much greater than the number of zeros it takes to write it. It takes only one zero to write 10, but the result is a number that is ten times greater than the number of zeros required to write it. In the same way it takes ten zeros to write 10,000,000,000, but the number written is ten billion, which is a billion times greater in size than the number of zeros used to write it.

Similarly it takes only ten decillion zeros to write the Earth-number, but the Earth-number itself is enormously greater than that number of zeros.

To write not ten decillion zeros, but an Earth-number of zeros, would require far more than the surfaces of all the objects in the known Universe, even with each zero the size of a hydrogen atom. A trillion such Universes as ours might suffice, and that is just to *write* Earth-number in a 1 followed by zeros. Skewes' number itself written by a 1 followed by an Earth-number of zeros, is *enormously*, ENORMOUSLY greater than the Earth-number that suffices to count those zeros.

So let's forget about counting zeros; that will get us nowhere. And if we abandon counting zeros, we don't need to have our exponents integers. Every number can be expressed as a power of ten, if we allow decimal exponents. For instance, by using a logarithum table, we can see that 34=10\1.53. So instead of writing Skewes' number as 10\10\10\34, we can write it as 10\10\10\10\1.53. (Such fractional exponents are almost always only approximate, however.)

There are some advantages to stretching out the large numbers into as many 10s as is required to make the rightmost number fall below 10. Then we can speak of a "single-ten number," a "double-ten number," a "triple-ten number," and so on. Skewes' number is a "quadruple-ten number."

We can't count objects and reach Skewes' number in any visual-izable way. We can't count zeros, either, and do it. Let us instead try to count permuations and combinations.

Let me give you an example. In the ordinary deck of cards used to play bridge, there are 52 different cards. (The number 52 is itself a "single-ten number" as are all the numbers between 10 and 10,000,000,000; $52=10\diagdown1.716$.)

In the game of bridge, each of four people is dealt thirteen cards. A player can, with equal probability, get any combination of thir-teen cards, and the order in which he gets them doesn't matter. He rearranges that order to suit himself. The total number of different hands he can get by receiving any 13 cards out of the 52 (and I won't bother you with how it is calculated) is about 635,000,000,000. Since this number is higher than ten billion, we can be sure it is beyond the "single-ten number" stage. Exponentially, it can be expressed as $6.35\times10\diagdown11$. Logarithms can help us remove that multiplier and put its value into the exponent at the cost of making that exponent a decimal. Thus $6.35\times10\diagdown11=10\diagdown11.80$. Since 11.80 is over ten, we can express that, exponentially, as $11.80=10\diagdown1.07$.

Consequently, we can say that the total number of different hands a single bridge player can hold is $10\diagdown10\diagdown1.07$. Using only 13 cards, we have, in a perfectly understandable way, reached a "double-ten number." We might almost feel that we were halfway to the "quad-ruple-ten number" that is Skewes'.

So let's take all 52 cards and let's arrange to have the order count as well as the nature of the cards. You begin with a deck in which the cards are in some certain order. You shuffle it and end with some different order. You shuffle it again and end with another different order. How many different orders are there? —And remember that any difference in order, however small, makes a different order. If two orders are identical except for the interchange of two adjacent cards, they are two different orders.

To answer that question, we figure that the first card can be any of the 52, the second any of the remaining 51, the third any of the remaining 50 and so on. The total number of different orders is $52\times51\times50\times\ldots\ldots4\times3\times2\times1$. In other words, the number of different orders is equal to the product of the first 52 numbers. This is called "factorial 52" and can be written "52!."

The value of 52! is, roughly, a 1 followed by 68 zeros; in other words, a hundred decillion decillion. (You are welcome to work out the multiplication if you doubt this, but if you try, please be prepared for a long haul.) This is an absolutely terrific number to get out of one ordinary deck of cards that most of us use constantly without any feeling of being overwhelmed. The number of different orders into which that ordinary deck can be placed is about ten times as great as all the subatomic particles in our entire Milky Way Galaxy.

It would certainly seem that if making use of 13 cards with order indifferent lifted us high up, making use of all 52 and letting order count will do much better still—until we try our exponential notation. The number of orders into which 52 different cards can be placed is $10 \backslash 68 = 10 \backslash 10 \backslash 1.83$.

That may strike you as strange. The number of orders of 52 cards is something like a trillion trillion decillion times higher than the number of bridge hands of 13 cards; yet while the latter is $10 \backslash 10 \backslash 1.07$, the former is only $10 \backslash 10 \backslash 1.83$. We're still in the "double-ten numbers" and we haven't even moved up much.

The trouble is that the more tens we add to such exponential numbers, the harder it is to move that rightmost component. For instance, a trillion is ten times as great as a hundred billion and counting a trillion objects would be an enormously greater task than counting a hundred billion. Write them exponentially, however, and it is $10 \backslash 12$ as compared with $10 \backslash 11$, and the rightmost components are only a unit apart. Write 12 and 11 as powers of 10 so that you can make use of "double-ten numbers" and a trillion becomes $10 \backslash 10 \backslash 1.08$, while a hundred billion is $10 \backslash 10 \backslash 1.04$ and the difference is scarcely noticeable.

Or put it another way. The number $10 \backslash 3$ (which is 1,000) is ten times as high as $10 \backslash 2$ (which is 100), but the degree to which $10 \backslash 10 \backslash 3$ is greater than $10 \backslash 10 \backslash 2$ would require a 1 followed by 900 zeros to be expressed. As for comparing $10 \backslash 10 \backslash 10 \backslash 3$ and $10 \backslash 10 \backslash 10 \backslash 2$, I leave that to you.

This is downheartening. Perhaps reaching the "quadruple-ten numbers" won't be that easy after all.

Let's try one more trick with 52 cards. Suppose each of the cards can be any card at all. Suppose the deck can have two tens of diamonds or three aces of clubs, or, for that matter 52 threes of hearts.

The total number of orders of such a chameleonlike deck could be calculated by imagining that the first card could be any one of 52, and the second card could be any one of 52 and so on for all 52. To calculate the number of different orders you would have to take the product of 52×52×52× . . . . 52×52×52; fifty-two 52s. This product which could be written 52\52 I might call "superfactorial 52", but if I do, I would be using a term I have just made up, so don't blame the mathematicians.

Superfactorials are immensely larger than factorials. Factorial 52 can be expressed by a 1 followed by 68 zeros; but superfactorial 52 is a 1 followed by 90 zeros, ten billion trillion times higher. Yet express it exponentially and superfactorial 52=10\90=10\10\ 1.95.

No good. We're still in the "double-ten numbers"

We'll just have to forget playing cards. We must have more than 52 units to play with, and we had better go all the way up; *all* the way up.

A generation or so ago, the British astronomer Arthur S. Eddington calculated that the total number of electrons, protons, and neutrons in the Universe was 10\79, or 10\10\1.90. This number is arrived at if we suppose that the Sun is an average star, that there are about a hundred billion stars in the average galaxy, and that there are a hundred billion galaxies in the Universe.

In addition to electrons, protons, and neutrons, of course, there are numbers of unstable particles unknown to Eddington, but their numbers are comparatively few. There are, however, massless particles such as neutrons, photons, and gravitons, which do not generally behave like particles but which are very numerous in the Universe.

If we wish, we can suppose that the number of massless particles speeding through space at any time is nine times the number of massed particles (probably a grievous overestimate) and make the total number of subatomic particles in the Universe 10\80, or 10\10\1.903.

Now, at least, we are starting with a "double-ten number" and that ought to do it. Skewes' number, here we come. All we have to do is take the superfactorial of 10\80, something we can express as (10\80)\(10\80).

Working that out (and I hope I'm doing it correctly), we get 10\10\81.9, or 10\10\10\1.91.

And that lifts us into the "triple-ten numbers" for the first time. In fact, if we compare the superfactorial of the total number of subatomic particles in the Universe, which is 10\10\10\1.91, and Skewes' number which, as a "triple-ten number," is 10\10\10\34, we might think we were almost there.

We need to begin with something more than the number of subatomic particles in the Universe—how about the amount of space in the Universe?

The smallest unit of space we can conveniently deal with is the volume of a neutron, a tiny globe that is about 10\—13 centimeters in diameter, or one ten-trillionth of a centimeter.

The Observable Universe has a radius of 12.5 billion light-years, or 1.25×10\10 light-years, and each light-year is equal to just under 10\13 kilometers. Hence, the Observable Universe has a radius of roughly 10\23 kilometers. Since 1 kilometer=100,000, or 10\5, centimeters, the Observable Universe has a radius of roughly 10\28 centimeters. From this we can calculate the volume of the Observable Universe to be roughly equal to 4.2×10\84 cubic centimeters.

A neutron with a diameter of 10\—13 centimeters, has a volume that is equal to roughly 5×10\—40 cubic centimeters. That means that the volume of the Observable Universe is roughly 2×10\124, or 10\124.3 times the volume of a single neutron.

Suppose we call the volume of space equal to that of a neutron a "vacuon." We can then say that there are 10\124.3 vacuons in the Universe and call that the "vacuon-number."

The vacuon-number is nearly a billion billion billion billion billion times greater than the number of subatomic particles in the Universe, so we can feel pretty confident about the superfactorial of the vacuon-number, which is (10\124.3)\(10\124.3), except that this comes out to 10\10\10\2.10.

Despite the vastly greater quantity of empty space than of matter in the Universe, the rightmost component of the "triple-ten number" went up only from 1.91 to 2.10, with 34 as the goal. That's enough to depress us, but wait—

In considering the number of vacuons in the Universe, we imagined it as existing at a moment in time. But time moves, and the Universe changes. A subatomic particle that occupies one place at one moment may occupy another place at another moment. The most rapidly moving particles are, of course, the massless ones which move at the speed of light.

The speed of light is just about $3\times10\backslash10$ centimeters per second, and the smallest distance one can move with some significance is the diameter of a neutron, which is $10\backslash-13$ centimeters. A photon will flash the width of a neutron, then, in about $3\times10\backslash-24$ seconds. We can consider this the smallest unit of time that has physical meaning and call it the "chronon."‡

For a long period of time, let's consider what we can call the "cosmic cycle," one period of expansion and contraction of the Universe (assuming it is oscillating). Some have guessed the length of the cosmic cycle to be 80,000,000,000, or $8\times10\backslash10$, years.

The number of chronons in one cosmic cycle, then, is roughly $10\backslash42$.

In every chronon of time, the Universe is slightly different from what it was in the preceding chronon or what it will be in the next chronon, because, if nothing else, every free-moving photon, neutrino, and graviton has shifted its position by the width of one neutron in some direction or other with each chronon that passes.

Therefore we might consider the total number of vacuons not only in the present Universe, but in the one that existed in the last chronon, the one that will exist in the next chronon, and, in general, all the Universes in all the chronons through a cosmic cycle. (To be sure, the expansion and contraction of the Universe alters its vacuon content, these increasing in number with expansion and decreasing with contraction, but we can suppose that the present size of the Universe is about average.)

In that case, then, the total number of vacuons through every chronon of the cosmic cycle is just about $10\backslash166.3$. What this means is that if you wish to place a proton somewhere in the Universe at some instant in time, you have (under the conditions I've described) a choice of $10\backslash166.3$ different positions.

But if you take the superfactorial of this enormous "total-vacuon number," you end up with $10\backslash10\backslash10\backslash2.27$.

We have hardly moved. I just can't seem to move those "triple-ten numbers" and make progress toward Skewes' number. I am Skewered.

In fact, it's worse than that. According to Mr. Boas, Skewes' de-

‡ Stanley G. Weinbaum once imagined space and time quantized in this fashion in one of his science fiction stories and used the word "chronon" for his ultimate particle of time.

termination of Skewes' number depended on the supposition that something called the "Riemann hypothesis" is true. It probably is, but no one has proved it to be so.

In 1955 Skewes published a paper in which he calculated the value of the number below which the number of primes *must* be higher at some point than the formula would predict, if the Riemann hypothesis were *not* true.

It turns out that the Riemann-hypothesis-*not*-true case yields a number that is far higher than Skewes' number. The new number, or what I suggest we call the Super-Skewes' number, is 10\10\10\ 1,000, or 10\10\10\10\3.

Super-Skewes' number and Skewes' number are both "quadruple-ten numbers"—10\10\10\10\3 and 10\10\10\10\1.53 respectively—and the difference in the rightmost component seems to be small. However, you saw what difficulty there was in budging the "triple-ten numbers" upward—well moving the "quadruple-ten numbers" upward is far harder still, and Skewes' number is virtually zero in comparison to Super-Skewes' number.

If I had reached Skewes' number, I would still have had Super-Skewes' number ahead of me. I would have been Super-Skewered.